符号中国 SIGNS OF CHINA

中国石

CHINA STONES

"符号中国"编写组 ◎ 编著

图书在版编目(CIP)数据

中国石：汉文、英文 /"符号中国"编写组编著. —北京：
中央民族大学出版社，2024.3
（符号中国）
ISBN 978-7-5660-2299-8

Ⅰ.①中… Ⅱ.①符… Ⅲ.①石—介绍—中国—汉、英 Ⅳ.①TS933

中国国家版本馆CIP数据核字（2024）第016899号

符号中国：中国石 CHINA STONES

编 著	"符号中国"编写组
策划编辑	沙 平
责任编辑	满福玺
英文指导	李瑞清
英文编辑	邱 械
美术编辑	曹 娜 郑亚超 洪 涛
出版发行	中央民族大学出版社
	北京市海淀区中关村南大街27号　邮编：100081
	电话：（010）68472815（发行部）　传真：（010）68933757（发行部）
	（010）68932218（总编室）　　　　（010）68932447（办公室）
经 销 者	全国各地新华书店
印 刷 厂	北京兴星伟业印刷有限公司
开 本	787 mm×1092 mm 1/16　印张：10.625
字 数	138千字
版 次	2024年3月第1版　2024年3月第1次印刷
书 号	ISBN 978-7-5660-2299-8
定 价	58.00元

版权所有　侵权必究

"符号中国"丛书编委会

唐兰东　巴哈提　杨国华　孟靖朝　赵秀琴

本册编写者

姚　琪

前言 Preface

欣赏与收藏奇石在中国源远流长，从秦汉时期出现，到唐宋元明清时盛行。千百年来，人们的采石、拣石、藏石、赏石、画石、论石之风久盛不衰，形成了博大精深的奇石文化。赏石成为中国人陶冶情操、与自然沟通的一种有效形式。历代文人墨客也曾为奇石所倾倒，留下不少佳话。

Appreciating and collecting wonder stones dates back to ancient times in China, starting from the Qin and Han dynasties, and increased momentum during the ensuing Tang, Song, Yuan, Ming and Qing dynasties periods. For thousands of years, the profound and extensive stone culture featured by the selection, collection, appreciation, painting and discussion of wonder stones prevailed in China. Appreciating stones has become one of the effective means for the Chinese people to cultivate their personality and to communicate with nature. Many poets and writers were overwhelmed by the appealing charm of wonder stones, and have left many beautiful verses and stories.

"Mountains without stones are not spectacular; rivers without stones are not clear; gardens without stones are not exquisite; and chambers without stones are not elegant." As a natural artwork, a piece of wonder stone is carefully shaped

"山无石不奇，水无石不清，园无石不秀，室无石不雅。"作为一种天然的艺术品，奇石经过大自然千百万年的鬼斧神工，石质、石形、石肌、石色、石韵等方面展现着特殊的美。本书介绍了石文化的传承，并且分门别类地介绍了各种奇石的成因、质地、文化历史、艺术特色、鉴藏知识等，并配以展示各类石材特征的相关图片，以文品石、以图赏石，从而让读者更加了解中国传统的石文化。

by nature through many years of uncanny workmanship, and it demonstrates its special beauty through quality, shape, texture, color and charm. While the book attempts to introduce the stone culture and heritage, it also discusses separately the formation, texture, cultural and historical, artistic features of different categories of wonder stones and the knowledge of wonder stone collection and appreciation. Pictures showing special characteristics of different types of wonder stones are included to supplement the text, so as to facilitate a better understanding of the traditional Chinese stone culture.

目录 Contents

中国石之奇
The Wonder of China Stones 001

中国石的历史源流
Historical Origins of China Stones 002

奇石的评估
Evaluating Wonder Stones ....,,.... 017

奇石的玩赏与陈设
Appreciation and Display of Wonder Stones ... 025

中国石之美
The Beauty of China Stones 031

造型石
Modeling Stones 032

纹理石
Striated Stones ....,,................... 074

矿物晶体
Mineral Crystals 116

古生物化石
Paleontological Fossils ...,,........................... 140

中国石之奇
The Wonder of China Stones

　　石头是自然界赋予人类的瑰宝，作为地球上最早的"居民"，石头构成了人类和其他一切生物的生活"舞台"。在人类的发展过程中，石头始终是文化的载体和传播者——从旧石器时代由天然石块充当简易的工具，到新石器时代的石斧、石刀的磨制；从营巢穴居时期简单以石头堆砌住处，到明清盛世允当典雅建筑的装饰材料；从远古的简单石制饰物，到后来的精美石雕和宝玉石工艺品——石头见证了人类文明的传承和进步。

Stones are "treasures" bestowed by nature to mankind as the earliest "inhabitants" on earth. Stones offered a "platform" for mankind and all other living creatures to live on. In the course of human development, stones have always served as the carrier and disseminator of culture. From simple tools made of natural stones during the Paleolithic Age to stone axes and blades during the Neolithic Age, from making simple shelters with stones in ancient times to using stones as elegant decoration materials during the Ming and Qing dynasties; from simple stone accessories in ancient times to exquisite stone carvings and stone artworks in modern times, stones have witnessed the continuation and progress of human civilization.

> 中国石的历史源流

人类诞生之后历经了漫长的石器时代。旧石器时代的标志是石器的制作，以一块石头打砸另一块石头，砸出锋利的薄刃，用来切

• 石英石砍砸器（旧石器时代）
A Quartz Chopper and Smasher (Paleolithic Age)

> Historical Origins of China Stones

Mankind went through a long period of Stone Age since born. Trademarks of the Paleolithic Age include the making of stoneware, using one piece of stone to chop and smash another to make sharp blades for cutting and stabbing. Trademarks of the Neolithic Age include the chiseled and abraded stone tools, which were more exquisite than the beaten and smashed stone tools. During that period, stone tools had a more specific division of purpose and a relatively more uniform shape for tools of the same category. Some had wooden handles affixed to facilitate the users. Compared to ancient times when stone tools were mostly used for hunting purposes and were easily processed and readily available, stones played an increasingly diverse role during the later

割、刺杀。新石器时代的标志是石器工具为凿磨而成，工具比打砸器精致，用途更加细化，同类石器的形制较为统一，有的石器还被安上木柄。相对于早期多数用来猎取食物、易于加工且随处可见的石头，其在原始社会后期的功用则越来越多元化，如用粗大的石材垒筑房屋，将美丽的石块作为赏玩或装饰品……内涵丰富的石文化由此开端。

stage of the primitive society: large stones were used to build houses while beautiful stones were used for appreciation or decoration. The stone culture with rich connotation emerged since then.

- 刮削器（旧石器时代）
  A Stone Scraper (Paleolithic Age)

- 尖状器（旧石器时代）
  A Pointed Stone Apparatus (Paleolithic Age)

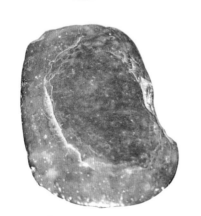

- 石斧（新石器时代）
  A Stone Axe (Neolithic Age)

- 石磨盘与石磨棒（新石器时代）
  A Stone Mill Disc and Rod (Neolithic Age)

石文化的发端

在先秦时代,虽然关于石的记载零星散见,描述也语焉不详,但作为石文化的先导和前奏,关于美石的采集、收藏活动已经出现了。

- 《阿房宫图》袁耀（清）

阿房宫是秦始皇（前259—前210）统一六国后在都城咸阳（今陕西省咸阳）修建的庞大宫苑,后毁于秦末的战火之中。

A Painting of Epang Palace by Yuan Yao (Qing Dynasty, 1616-1911)

Epang Palace was a huge and magnificent palace compound built by the first Emperor of the Qin Dynasty (259 B.C.-210 B.C.) in Xianyang (current day Xianyang City, Shaanxi Province), capital city of the Qin Dynasty after he conquered the Six States and unified the country. The Palace was destroyed during the war in late Qin Dynasty.

Emergence of the Stone Culture

Although there were only a few scattered written records about stones and the descriptions were by no means detailed, as the forerunner and prelude to the stone culture, activities to collect beautiful stones emerged during the pre-Qin times.

The earliest written record of beautiful stones in China appeared in an ancient book entitled *The Classics of Mountains and Rivers*. In chapter five of the book, there were several dozens of descriptions about stones. In another Warring State Period classic *Biography of Emperor Mu*, introductions to beautiful stones were found in this fabled book of mythical stories. It reads: "The King toured Wenshan Mountain for three days, where he collected colorful stones." King Zhoumu used these beautiful stones on grand occasions such as offerings at memorial services and prizes for awarding officials.

Another book written later than *The Classics of Mountains and Rivers* named *Book of History, Yu Gong*, was regarded as the most scientific geographical book during the pre-Qin times. It is specified in the book that there were "strange stones" in the city of Qingzhou, and "musical stones" in the city of Xuzhou.

中国关于美石的最早记录见于古籍《山海经》，其五卷《山经》中关于石的记载有几十条之多。战国时期成书的《穆天子传》，是一部充满想象的神话典籍，其中也有关于美石的介绍："天子三日游于文山，于是取采石。""采"是彩色之意。采得彩色石的周穆王用其祭奠、分封，场面十分隆重。

成书晚于《山海经》的《尚书·禹贡》被认为是先秦最富于科学性的地理记载，其中记载青州有"怪石"、徐州有"泗滨浮磬"。

Both stones were appreciated for their incredible texture, sound and color and were treated as "tribute items".

During the Qin and Han dynasties, appreciation stones were confined mostly to imperial gardens and noble homes. As described in many classic books and poems during the Qin and Han dynasties, there were many scenic spots in imperial gardens that were decorated with wonder stones. For instance, the Efang Palace built by the first Emperor of the Qin Dynasty (Qinshihuang), the Weiyang Palace and Shanglin Gardens of the Han Dynasty were all such gardens. The "Liang Garden", a private garden

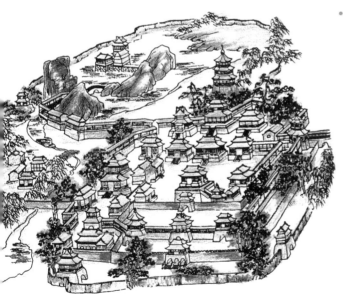

● **西汉建章宫复原图**
建章宫是西汉武帝刘彻于太初元年（前104）建造的宫苑。建章宫的西北部开有大池，名太液池，是一片以园林为主的区域。池中用土石堆筑了三个岛屿，象征传说中的瀛洲、蓬莱、方丈三座山。

Restored Painting of the Jianzhang Palace of the Western Han Dynasty（206 B.C.-25 A.D.）
Jianzhang Palace was built by Emperor Wu (Liu Che) in the first year of the Taichu period (104 B.C.) of the Western Han Dynasty. At the northwest corner of Jianzhang Palace, a big lake named Taiye was built. The area was predominantly gardens and trees. In the lake, three small islets were built with soil and gravel, symbolizing the three fabled mountains of Yingzhou, Penglai and Fangzhang respectively.

因石质、声色的不同，被作为赏玩之物而列为"贡品"。

秦汉时期，赏石仍局限于皇家、贵族的园林。从秦汉时代的古籍和诗文所描述的情景得知，秦始皇的阿房宫，汉代的未央宫、上林苑所用观赏性奇石点缀的景点颇多。东汉大将军梁冀的私人宅苑"梁园"中也收罗有大量奇珍怪石。此时赏石文化在造园实践中得到了较大的发展，置石造景、寄情物外。

隋唐时期的赏石之风

从魏晋到隋唐时期，赏石摆脱了园林假山的局限，逐渐成为独立欣赏的对象，从而形成了收藏、鉴赏室内观赏石的风气。此时具有艺术气质的官宦与文人雅士成为赏石的主流，将"小而奇巧者"作为案头清供，复以诗记之，以文颂之，从而使天然奇石的欣赏具有浓厚的人文色彩，开创了中国赏石文化的一个新时代。唐代一些著名诗人，如王维、杜甫、刘禹锡、白居易都是爱石之人，并写过咏赞奇石的诗歌。白居易曾在杭州任地方长官，卸任时别无长物，只带走了数块石

of General Liang Ji of the Eastern Han Dynasty, collected many pieces of spectacular wonder stones. During that period, the stone appreciation culture enjoyed rapid development in garden building. Stones were used to create scenic wonders and to express sentiments.

Trends of Stone Appreciation during the Sui and Tang Dynasties

During the Wei, Jin, Sui and Tang dynasties, appreciation stones broke through the limitations of garden rockery, and gradually became independent appreciation objects. A culture of stone collection and appreciation emerged. In those days, officials with artistic temperaments and literati comprised the mainstream crowd in stone appreciation. They put stones that were "small and exquisite" on their desks to enjoy and worship, and wrote poems and lyrics to praise them. Consequently, the appreciation of natural wonder stones reflected deep humanities culture and a new era of stone appreciation culture in China was initiated. Some famous poets of the Tang Dynasty such as Wang Wei, Du Fu, Liu Yuxi and Bai Juyi, were all lovers of wonder stones. They wrote many poems to praise these beautiful

头,他作诗提到此事:"三年为刺史,饮冰复食蘖。唯向天竺山,取得两片石。此抵有千金,无乃伤清白。"此外白居易还有许多赏石诗文,他的《太湖石记》更是反映唐代赏石盛况及文化水准的代表作之一。

objects. One of the famous poets Bai Juyi who had served as a local official in the city of Hangzhou for many years, when he completed his term of service and left Hangzhou, he took with him nothing but some pieces of wonder stones. He mentioned this in one of his poems: "During the three years when I served as the governor of Hangzhou, I led a simple and frugal life. Fortunately, I got two pieces of wonder stones from the Tianzhu Mountain. For me, they are more valuable than money. I hope this act will not spoil my reputation of being an honest official." In addition, he also wrote many poems and essays to praise the stones. His article *Records of Taihu Stones* was one of the works reflecting the flourishing stone appreciation culture during the Tang Dynasty.

- 玲珑(太湖石)

  唐代诗人白居易曾在《太湖石记》中称赞太湖石:"三山五岳,百洞千壑,覙缕簇缩,尽在其中。百仞一拳,千里一瞬,坐而得之。"意即:三山五岳,百洞千壑,弯弯曲曲,丛聚集缩,尽在其中。自然界的百仞高山,一块小石就可以代表;千里景色,一瞬之间就可以看过来。这些坐在家里就能享受到的。

  Exquisite Stone (Taihu Stone)

  In his article *Records of Taihu Stones*, Bai Juyi, a famous poet of the Tang Dynasty, praised the Taihu Stones as miniatures of famous mountains with hundreds of caves and gullies. A small piece of Taihu Stone is a miniature of a high mountain in nature; a landscape of one thousand miles can be browsed through between moments. All these can be enjoyed sitting at one's home.

陶渊明与醒石

魏晋南北朝时期，奇石开始成为人们独立欣赏的对象。东晋大诗人陶渊明喜欢与石为伴，留下一段佳话。相传，陶渊明住宅旁边的菊丛之中，有一块平滑的大石，他每逢贪杯喝醉了，便坐卧其上，而且诗兴大发，写下一首首耐人寻味的诗歌。陶渊明认为这块大石能让他醒酒，又能让他诗思泉涌，于是便给其起名"醒石"。陶渊明与醒石结缘，引起后世文人浓厚的兴趣和遐思。明代林有麟所著的《素园石谱》是中国第一部图文并茂的石谱，他在这本书中将"醒石"的形状形象地描绘出来。也因为这一段佳话，陶渊明被后人尊奉为"赏石祖师"。

Tao Yuanming and the Sober Stone

During the Wei, Jin, Southern and Northern dynasties, wonder stones became independent appreciation objects in China. Tao Yuanming, a great poet of the Eastern Jin Dynasty, was fond of stones and had left some interesting anecdotes about them. It was said that there was a big flat stone among a tuft of chrysanthemums beside his house. When Tao Yuanming got drunk, he would sit on it and his poetic reveries would be inspired and he could then write many lines of thought-provoking poems. Tao thought since this big stone could wake him and trigger his poetic reveries, so he named it the "Sober Stone". The intimate relationship between Tao and his sober stone elicited great interests from many young literati. Lin Youlin of the Ming Dynasty wrote a book titled *Scholars' Rocks in Ancient China*, which is the first map of stones with both texts and illustrations in China. It was in this book that the Sober Stone was vividly described. For this reason, Tao Yuanming was honored as the "founding father" of the stone appreciation culture by descending generations.

• 《渊明醉归图》张鹏（明）
Tao Yuanming Returning Home Drunk by Zhang Peng (Ming Dynasty, 1368-1644)

宋元时期的赏石文化

宋元时期赏石文化蓬勃发展，上层社会形成爱石、藏石的浓厚风气，文人雅士的参与，使赏石作为一种艺术融入书画殿堂。

北宋大文学家苏轼同时也是一位富于传奇色彩的赏石家。他与奇石、砚石有着特别的情缘，曾收藏有仇池石、壶中九华、沉香石、雪浪石等名石。他在定州任官时的书屋"雪浪斋"，就是取名于雪浪石。他还曾写过咏雪浪石的七言长诗："画师争摹雪浪势，天工不见雷斧痕……此身自幻孰非梦，故园山水聊心存。"苏轼长期的赏石经历，使他的藏石观与书法绘画理论相互融合、借鉴。

同时代的书画家米芾对苏轼所画的木石极为赞赏，称"子瞻（苏轼的字）作枯木，枝干虬屈无端。石皴硬，亦怪怪奇奇无端，如其胸中盘郁也"。而米芾本人更是堪称中国宋代最有名的藏石、赏石大家。米芾因爱石成癖，对石下拜而被时人称为"米癫"。元人有《题米南宫拜石图》诗："元章爱砚复爱石，探瑰抉奇久为癖。石兄足拜

Stone Appreciation Culture during the Song and Yuan Dynasties

The culture of stone appreciation enjoyed vigorous development during the Song and Yuan dynasties. A strong trend for stone appreciation and collection was formed among the upper class. Thanks to advocacy efforts by many scholars and literati, stone appreciation was admitted by the painting and calligraphy communities as a form of art.

Su Shi, a famous writer of the Northern Song Dynasty, was also a legendary connoisseur of wonder stones. He was especially fond of ink-stones and other rare stones. He collected many pieces of precious wonder stones including the Qiuchi Lake Stone, Mount Jiuhua in a Pot, Incense Stone and Snow-wave Stone. He named his study in Dingzhou City where he served as the governor the "Snow-wave Hall". He once wrote a long seven-character poem to praise the Snow-wave Stone. Some of the lines read as follows: "Painters strive to depict the momentum of the snow wave, with such fine craftsmanship only found in heaven … This is not a dream. I can envision and enjoy mountains and rivers in the stones." His long-time

自写图，乃知癫名传不虚。"可见米芾在赏石过程中常能获得书画创作的灵感。同时，米芾还能将自己的书画创作理论用于品石，其在相石方面创立的"瘦、皱、漏、透"四字原则至今被奉为圭臬。

北宋徽宗皇帝赵佶是历史上少有的艺术帝王，他酷爱山石，对奇石有独到的鉴赏力。他为了营造宫苑"艮岳"，动用了上千艘船只为专门从江南运送山石花木。这就是宋史上著名的劳民伤财的"花石纲"，直接导致了方腊领导的农民起义。元人郝经有诗叹曰："中原

● 宋徽宗像
Portrait of Emperor Huizong of the Song Dynasty

stone appreciation experience enabled him to incorporate his outlook in stone appreciation in his calligraphy and painting theories.

Mi Fu, Su Shi's contemporary painter of the Northern Song Dynasty, was very fond of Su Shi's paintings of trees and stones. He praised that "in Su Shi's paintings, withered trees have endless branches. The stones are full of wrinkles, hard, grotesque and without a clue, like his frustration." Besides, Mi Fu himself was the most famous artist in stone collection and appreciation of the Song Dynasty. His love for stones was so strong that it became his obsession. He even worshipped stones. Hence he got the nickname "Crazy Mi". A Yuan Dynasty poet inscribed a poem on the painting "*Mi Nangong Worshipping Stones*", which reads: "Mi loves both ink-stones and wonder stones. He is addicted to collecting wonder stones. The painting shows Mi Fu bowing to the stone, testifying to his well-deserved the reputation of 'Crazy Mi'." Evidently, we could see that Mi Fu did draw inspiration from appreciating stones in his calligraphy and paintings. He could also apply his theories on calligraphy and painting in stone appreciation. The

• 北京北海琼岛上的假山石

琼岛又名琼华岛，始建于金大定六年（1166）。"琼华"意为美玉如花，形容岛上满山堆积的太湖石。这些太湖石来自北宋故都开封的皇家园林"艮岳"。岛上被艮岳奇石覆盖，远望就像是堆满了白云，因此这处景观在明代时被命名为"琼岛春云"。

Rockeries on the Jade Islet in Beihai Park, Beijing

Jade Islet, also known as the Qionghua Islet, was built in the sixth year of the Dading Period of the Jin Dynasty (1166). "Qionghua" means jade as beautiful as flowers. The term was used to describe the Taihu Stones covering the islet. These stones were shipped from the imperial "Genyun Garden" in Kaifeng, former capital city of the Northern Song Dynasty. Because the islet is covered by piles of wonder stones, looking from afar, they are like overlapping clouds. This scenic spot was named "Spring Clouds over Jade Islet" during the Ming Dynasty.

自古多亡国，亡宋谁知是石头。"嗜石误国，赵佶终成为亡国之君。随着金兵南下，东京陷落，艮岳之石散落，迁移各地，今人难以得

four-letter criteria of "slender, wrinkled, hollow and transparent" initiated by Mi Fu is still regarded as the guiding principle in evaluating wonder stones.

Emperor Huizong of the Northern Song Dynasty (Zhao Ji) was one of the few emperors with an artistic temperament in Chinese history. His vehement love for mountain stones nurtured his unique taste for stone appreciation. In order to build his garden Genyue, he ordered hundreds of boats to ship beautiful stones, flowers and trees from southern China. This is the notorious "Huashigang Event", which directly fused the peasant uprising led by Fang La. Hao Jing, a poet of the Yuan Dynasty lamented in one of his poems, "There were many causes for the demise of states in Central China. Surprisingly, it was the stone that led to the death of the Song Dynasty." Because of his obsessed love for stones, he ignored his duties as the head of state. Zhao Ji was finally defeated. With the fall of the capital city Dongjing, Song's territory was occupied by the Jin army from the north. Stones collected by Emperor Zhao Ji for his Genyue Garden were dispersed and hardly seen nowadays. Thanks to painters of the time who sketched shapes of those

见。所幸这些奇石的形态被当时的画家用线条描摹下来，使后人多少可以了解艮岳之奇壮。

wonder stones, we now could have a glimpse of the grandeur of the Genyue Garden.

米芾爱石成痴

宋代著名的书画家米芾，非常喜好奇石。据传他刚到一处官衙上任，看见立在府内的奇石独特，一时心喜，便让随从给他拿来上朝用的袍笏和官服，穿戴整齐后对着奇石行叩拜之礼。还有一次，他外出时见到一块奇石，欣喜若狂，便绕石三天，搭棚观赏，不忍离去。后人在他搭棚拜石处修建了一座"拜石亭"，还在奇石与亭子之间修建了"绕石桥"。他在江苏为官时，因为当地毗邻盛产美石的安徽灵璧县，便常去搜集奇石，回来后终日把玩，闭门不出，影响了政务。有一次，上级官吏杨杰到米芾任所视察，得知此事，便对米芾说："朝廷把千里郡邑交给你管辖，你怎么能够整天玩石头而不管郡邑大事呢？"谁想米芾却从袖中取出一枚清润玲珑的灵璧石，一边拿在手中反复把玩，一边对杨杰说："如此美石，怎么能不令人喜爱？"杨杰未予理睬。米芾又从袖中取出一枚更加奇巧的灵璧石，如此一而再，再而三，杨杰实在无法抵挡诱惑，终于开口说道："难道只有你喜欢？我也非常喜爱奇石。"说着他一把将米芾手中奇巧的灵璧石夺了过去，竟忘记了此行巡察的目的，心花怒放地回去了。

•《米芾拜石图》陈洪绶（明）
Mi Fu Paying Tribute to the Stones by Mr. Chen Hongshou (Ming Dynasty, 1368-1644)

Mi Fu's obsessed love for stones

Mi Fu, a famous calligrapher and painter of the Northern Song Dynasty, was deeply in love with wonder stones. Once he was assigned to work in a new city where he discovered a piece of unique wonder stone in the government office. He was so delighted, so he ordered the staff to bring his official costume and hand panel, and bowed down to the wonder stone. One other day when he was on an official travel, he saw a piece of wonder stone on the way and got so excited. He stopped and ordered a simple shed be built near the stone. He then walked around the stone for three days to observe and enjoy it. Later, people built a "Stone Worshipping Pavilion" at the spot of the shed and a "Stone Circling Bridge" leading from the pavilion to the stone. When Mi Fu was assigned to work in Jiangsu Province, he often went to the neighboring Lingbi County, Anhui Province, where rich resources of wonder stones existed. There he would look for beautiful wonder stones. When he returned home, he would shut himself up in his room to study and enjoy the stones. This affected his work adversely. One day, his supervisor Yang Jie came to the city to inspect his work. Yang Jie said, "The government appointed you as the administrator of the county. How can you overlook your duties by enjoying the stones only?" Mi Fu took out a beautiful stone from his sleeve and said, "How could anyone stop loving such a beautiful piece of art!" Yang Jie did not respond. Mi Fu took out a piece of Lingbi stone, which was more exquisite and more elegant. After several rounds of showing, Yang Jie could not remain silent any more. He said, "You are not the only one who loves stones. I also do." Having said that, he grasped the Lingbi stone from Mi Fu and left happily, forgetting completely the mandate of his mission.

明清时期的赏石理论

明清两代是中国古代观赏石文化的全盛时期，品石专著层出不穷，流传至今的石谱、石录、石赞等有数十种之多，此外还有大量砚石、印石著录。在这些赏石专著或专论中，有的记载了奇石的产地、特征，有的介绍了奇石布置、保养的心得，有的记述了对藏石的礼赞

Stone Appreciation Theories during the Ming and Qing Dynasties

The Ming and Qing dynasties witnessed a flourishing period of the stone appreciation culture in China. Many monographs on stone appreciation were published and about several dozens of such books including Map of Stones, Catalogue of Stones and Praise of Stones survived until now. In addition, there

题名，有的讲述了得石的艰辛和收藏后的喜悦，更有的图文并茂，录有名人诗文、掌故，并绘有石头的线条图，直观性极强。其中颇具代表性的有明代计成的《园冶》、林有麟的《素园石谱》等。计成是明代著名造园大师，所著《园冶》开造园艺术的理论和工艺见诸笔墨之先河，特别强调奇石在中国园林中的特殊位置，讲求石形的奇妙。明万历年间林有麟的《素园石谱》图文并茂，长达四卷，对明代的赏石理论与实践进行了全面的概括。他在《素园石谱》中介绍了"石近于禅"的境界，把赏石的意境提升到

were many works and catalogues on inkstones and chops. Some of the books described the areas where the stones were produced and their characteristics. Others recorded experiences in the layout and maintenance of wonder stones. Still others included collections of poems and stories on stone collection and appreciation. Some even described the hardship while searching for stones and the joy after obtaining them. Many books were enriched with illustrations, poems by famous literati, interesting anecdotes, and line charts of stones, making the visual effects very impressive. Among them, the book entitled *Garden Building* by Ji Cheng and *Scholars' Rocks in Ancient China* by Lin Youlin of the Ming Dynasty were representatives. Ji Cheng was a famous garden architect of the Ming Dynasty. His book *Garden Building* was the first to put in writing the theory and practice of garden building. The book emphasized the unique status of wonder stones in traditional Chinese garden buildings and the importance of stone shapes. During the Wanli period of the Ming Dynasty (1573-1620), Lin Youlin compiled a book *Scholars' Rocks in Ancient China*. It is a four volume serial containing many illustrations. It gives a thorough review

• 墨彩竹石纹六角瓷花盆（清）
Hexagonal Ceramic Flowerpot with Bamboo and Stone Patterns (Qing Dynasty, 1616-1911)

- 苏州环秀山庄内的假山石

环秀山庄占地不大，景观以湖石假山为主。园内假山是由清代叠山大师戈裕良所堆，不仅峭壁、峰峦、洞壑、涧谷、磴道等应有尽有，而且极富变化。游人入园如置身于万山之中，因而有"别开生面、独步江南"之誉。

Rockeries in Huanxiu Garden, Suzhou

Huanxiu Garden occupies a small area and primarily composed of lakeside rocks and rockeries. These artificial hills were built by Master Ge Yuliang of the Qing Dynasty. They were not only built into various crags, apexes, caves, gullies and paths, but also with lots of variation. When visitors came to the garden, they felt as if they were in a thousand hills. The garden hence won the reputation of being a "ground-breaking and fresh-looking garden in southern China".

了具有人生哲理、内涵更为丰富的哲学高度。这是中国古代赏石理论的一次飞跃。

清代著名画家郑板桥完善了宋人的赏石观。他认为米芾的"瘦、皱、漏、透"虽"尽石之妙"，"而不知陋劣中有至好"，石中丑而雄、丑而秀者亦是佳品，怪石丑到

and summary of the stone appreciation theory and practice during the Ming Dynasty. It also introduced the concept of "stones are closer to meditation", uplifting stone appreciation to the philosophical level with rich living principles and connotation. This marked a leap forward of traditional Chinese stone appreciation theory.

极处，便是美到极处，道出了丑石观的真谛。

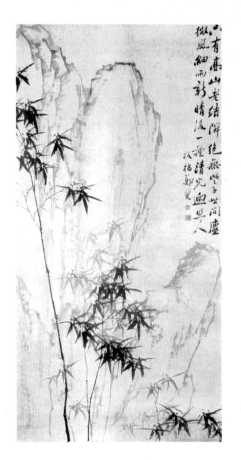

《竹石图》郑板桥（清）
Bamboos and Stones by Zheng Baiqiao (Qing Dynasty, 1616-1911)

Zheng Banqiao, a famous painter of the Qing Dynasty, improved upon the stone appreciation theory over that of the Song Dynasty. He declared that although Mi Fu's criteria of "slender, wrinkled, transparent and hollow" fully expounded the "beauty of stones", but Mi Fu did not realize that "the best beauty could also be concealed in ugly stones". He argued that ugly but majestic, ugly but delicate stones could also be excellent pieces. An extremely ugly-looking stone is also an extremely beautiful one. This explains the essence of Zheng Banqiao's outlook on ugly stones.

> 奇石的评估

奇石是靠大自然千百年来的冲刷、打磨而成，是集优异的质地、形态、色泽、纹理于一体的不可多得的观赏品，天然的神秘性是其突出特点。对一件奇石的综合评估常从六个方面入手，依次为质地、形态、色泽、纹理、声音、意蕴。

质地

石质指构成石品的矿物成分，能体现石的灵气，而灵气则具有观赏和使用价值。石质包括硬度、密度、质感、光泽等因素。其中，硬度是决定石质优劣的关键。一般以硬度来判断石质优劣，石品的密度越大，硬度越高。石质光滑细腻、质地纯净、硬度高、光润透明者为

> Evaluating Wonder Stones

Wonder stones have been washed and polished by nature for thousands of years. They are precious appreciation objects with integrated characteristics including excellent texture, shape, color and grain. An outstanding feature of a wonder stone is its natural mystery. A piece of wonder stone is usually evaluated from the following six aspects in the order of: texture, shape, color, grain, sound and charm.

Texture

A stone's texture refers to the mineral ingredients comprising the rock which can embodies the soul of a stone. As the soul of the stone, it determines the stone's value for appreciation and utility. The stone's texture is usually determined

• 质之中藏（黄河源头石）
A Top Grade Stone (A Stone from the Headstream of the Yellow River)

上等；结构松软、陈杂、表层粗糙、灰暗、硬度低者次之。但石种不同，质地也会有所不同，不能强求一律。

形态

对于某些石头，应多从形态、体量、大小、轮廓等方面评估。石之"美"的形态并无统一标准，有的以奇形怪状、瘦削玲珑的形体展示美感，有的则以浑圆古朴、粗犷憨实的形体展示美感，更有的以状物象形、惟妙惟肖而展示美感。

北宋书画家米芾以"瘦、皱、漏、透"四个字高度概括了中国传

by several factors such as its hardness grade, density, appearance and color etc., Among these factors, hardness grade is the key. Generally speaking, higher density stones also have a harder texture. A stone with a fine and smooth surface, pure texture, harder and more transparent is of superior quality while a stone with a loose structure, rough surface, soft and dull color is of secondary quality. However, as different types of stones have different textures, it is inappropriate to apply unified criteria.

Shape

Some of the stones are evaluated in terms of the shape, size, appearance and profile. There are no unified criteria to measure the "beauty" of a stone. Some stones show their beauty by displaying their grotesque shape, slender and exquisite body. Others do it by displaying their plain, voluminous and rustic body. Still, others show their beauty by mimicking other creatures.

The four letter criteria "slender, wrinkled, hollow and transparent", first introduced by Mi Fu, a famous painter and calligrapher of the Northern Song Dynasty, have accurately summarized the traditional Chinese concept

统赏石理念和审美标准。观赏石的"瘦"是指石体修长。北宋时人重清秀、瘦美，因此清瘦可以说是古代文人的象征。现在留存在苏州留园的著名古石"冠云峰"是"瘦"的典型。它修长挺拔而中间有束腰变化，线条流畅而有力，给人一种高耸入云的壮美感受。"皱"是表面有凹凸变化的褶皱，它是由地壳运动和风化剥蚀形成的，是岁月的

and aesthetical standards on stone appreciation. "Slender" means a stone is thin and tall. People during the Northern Song Dynasty admired bony beauty, which was taken as the typical image of literati and scholars in ancient China. A typical example of a "slender stone" is the "Cloud Capped Peak", a famous piece of old stone kept in the Lingering Garden, Suzhou City. The stone is tall and slim with recessed waist lines in the middle

• 苏州沧浪亭内的太湖石
Taihu Stone in the Surging Wave Pavilion, Suzhou

• 北京恭王府花园的独乐峰
The Lonely Joy Peak in the Garden of Prince Gong's Mansion, Beijing

留痕，因此给人以厚重的沧桑感。中国古代文人欣赏奇石，面对石头表面的皱褶，不免想到历史变迁、人生易老、仕途艰险而发出感叹。

"漏"和"透"是指石体中出现的前后、上下贯穿的洞穴，它们是在地质运动中形成的。大小不等、方向各异的洞穴组成匪夷所思的奇妙结构。"漏"具有幽深感和神秘感，"透"又具有通明感和空灵感。一块奇石往往同时具备"瘦、皱、漏、透"四种品质，给人以玲珑剔透、婀娜多姿的美感。

色泽

石头的色泽一般要求典雅，艳而不俗。有些奇石，如蜡石等的自

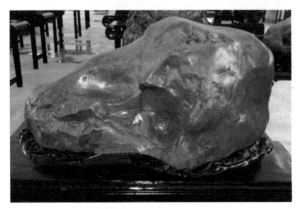

● 黄蜡石
Yellow Wax Stone

and the drapery lines are smooth and dynamic, rendering a sensational feeling of magnificent beauty. "Wrinkled" refers to uneven stone surfaces dotted with convex and concaves that were results of the crust movement and wind erosion in ancient times. As these wrinkles are traces of time, they convey a strong message of bitterness. Facing these wrinkles on stones, Chinese scholars could not help reflecting on historical changes, life's vulnerability and the challenges in pursuing a career. "Hollow" and "transparent" refer to caves and burrows in and around the stone body. As these holes were formed during crust movements, their structures are incredibly intricate. "Hollow" gives a sense of serenity and mystery while "transparent" gives a sense of space and openness. A wonder stone possessing all four features of "slender, wrinkled, hollow and transparent" usually impresses the viewer with its exquisite and graceful beauty.

Color

A stone's color has to be elegant, dazzling but not vulgar. Some wonder stones, such as decoration pieces of natural alabaster, have very high and strict requirements on color. Differences of stones' colors,

然形摆件，对色泽要求十分严格。石头色泽上的差异，在价值上的差异巨大。有些奇石呈现出多种色泽的交混，斑斑块块形成优美图案，也是奇石的价值所在。

各个石种有不同的要求。如昆石、钟乳石以晶莹、雪白为上，黄蜡石以纯黄凝冻为上，太湖石以青白为上，灵璧石以玄黑为上；卵石类中也有很多属于色彩石。色泽单纯或多重色彩巧妙搭配的奇石均可能归入上品。一般来说，具象石类与抽象石类的色彩以沉厚古朴的深色系列为佳，最忌颜色的混浊不清和刺激性的"俏"色。

纹理

指石品上的圆、点、线、面及块状等所构成的纹路图案或形状。奇石上的花纹分凹、凸纹，平、斜纹，点、线纹，粗、细纹，面、块纹；纹又分单色、双色、混合色纹。凡线条节奏明快、纹理富有韵律、变化无穷、妙趣横生的美石，都可称为精品石。

纹理是石头经亿万年自然风化的结果。有些石纹不仅千姿百态、变

reflected in values, are huge. Some stones have mixed colors, and they form beautiful patterns. This also contributes to the stone's high value.

Different types of stones have different grades on color. For example, the best color for Kun Stone and Stalactite is crystal white while the best color for Yellow Wax Stones is pure and congealed yellow. Taihu Stones take greenish white as the top grade while Lingbi Stones take dark black as the top grade. Many cobble stones are also colorful. In short, wonder stones of pure

• 圣火（金沙江石）
The Holy Flame (Stones from the Jinsha River)

化万千，而且与石形、石质、石色相结合，为人们赏石提供了无穷的乐趣。甚至有的石头纹理具有拟人、拟物、拟文字等装饰意象而成为绝品。南朝刘勰在《文心雕龙》中说："云霞雕色，有逾画工之妙；草木

color or of wisely mixed colors can all be classified as top grade stones. Generally speaking, for concrete or abstract stone categories, darker color series are more admired than turbid colors or excitant "bright" colors that are usually tabooed.

Grains

Grains refer to patterns or shapes formed by circles, dots, lines and blocks on the stone. Grains on wonder stones can be divided into different categories including concave and convex grains, horizontal and slanting grains, dotted and lined grains, bold and fine lines, full or partial grains; grains can be further divided into single-color, dual-color and mixed-color

- 函关负隐（彩陶石）

"函关负隐"来自春秋时期道家学说的创始人老子的典故，据说他在春秋末年骑青牛来到函谷关（今河南省灵宝市北），留下五千言的道家经典《道德经》，之后出关西去，不知所终。这块彩陶石表现了老子骑青牛的形象。

Ridding at the Hangu Pass (Color Pottery Stone)

The name of the painting "Ridding at the Hangu Pass" was inspired by the story about Lao Zi, a philosopher during the Spring and Autumn period and founder of the Taoist theory. The story goes as follows: Lao Zi rode a green buffalo and came to Hangu Pass (located in Lingbao City, Henan Province). There he wrote the famous five-thousand character scripture on "Tao Te Ching" of Daoism. Then he went out of the pass to the west and disappeared. This piece of stone represents the image of Lao Zi riding on the back of a green buffalo.

- 敦煌壁画（黄河石）
Dunhuang Fresco (Yellow River Stone)

贲华，无待锦匠之奇。夫岂外饰，盖自然耳。"意思是说，云霞变换设色，超过了画家的绘画；草木彩花，织锦的工匠也织不出来，这一切都不是外加的装饰，而是自然生成的。

声音

有些奇石在敲击后会发出悦耳的声音，如灵璧石，早在先秦时代就被用来制作乐器。但这只是某些奇石的特性，故其审美价值、收藏价值也会相应增加。

• 凤鸣灵山（灵璧石）
Phoenix Singing at Mount Ling (Lingbi Stone)

types. Beautiful stones with clear lines, rhythmic grains and infinite variations can all be deemed excellent pieces.

Grains have resulted from billions of years of natural weathering. Gains on some stones are not only vivid and varied, but also integrate well with the stone's shape, texture and color. Such stones usually offer the viewer infinite joy. In particular, grains on some stones carry decorative images that mimic humans, creatures, characters or words. Such stones are regarded as unique pieces. In his book *Carving a Dragon at the Core of literature*, Liu Xie, a famous scholar of the Southern dynasties said, "The cloud has its natural color; they look more vivid than the cloud in the picture painted by the masters. Grain patterns of flowers and trees cannot be woven by craftsmen. All these beautiful scenes are not artificial decorations; rather they are works of nature. "

Sound

Some wonder stones can produce dulcet sounds when knocked. For example, the Lingbi Stone was used to make musical instruments as early as during the pre-Qin times. As this special feature only exists in some wonder

意蕴

奇石所表现出来的意境、气韵、意蕴，是奇石之形、质、色、纹、声的综合表现。中国的赏石文化受儒家思想影响，主张以德赏石、赏石修德，石美人更纯；提倡通过赏石，学习石之表里如一、质朴诚实、坚韧宽厚、纯洁坦荡等自然品德，以此修炼自我。此外，道家思想崇尚自然、追求意境，反对雕琢堆砌、矫揉造作，也被后人运用到奇石的陈设、养护等实践之中。

stones, their aesthetical and collection values increase accordingly.

Charm

A stone's charm refers to the artistic ambience and the implication the stone is impregnated. It usually involves the stone's shape, texture, color, grain and sound. Stone appreciation culture in China has been heavily influenced by Confucianism, which maintains that stone appreciation should be pursued in an ethical approach, and should help cultivate ethics. Further, it should achieve a state of harmony between beautiful stones and noble-minded persons. The stone appreciation culture also calls for learning from the natural virtues of stones such as their uprightness, honesty, integrity, perseverance, generosity and purity. In addition, Taoism principles such as the admiration for nature, pursuit of implication and opposition to polishing, piling and affectation have been applied in the display and maintenance of wonder stones by later generations.

- 秋之韵（乌江石）
The Rhythm of Autumn (Wujiang Stone)

> 奇石的玩赏与陈设

形态各异、玲珑剔透、纹理奇特、色泽秀丽的石头,或陈设于案头茶几,或布置于庭院,或点缀于园林,可谓"花不解语还多事,石不能言最可人"。

叠山置景

"无石不成园",石头是中国古典园林中最基本的造园要素之一,因其具备了象外之象、景外之景的生发能力,在园境营造中发挥着不可替代的独特作用。中国古代造园通过对石头的巧妙利用和设置,体现出中国园林独特的山水自然情趣,也营造出了独具特色的园林意境。

石是叠山的材料,山是中国古典园林的骨架,是园景营造的重

> Appreciation and Display of Wonder Stones

Stones of exquisite shapes, unique grains and elegant colors can be set on desks and tea tables for appreciation or in courtyards and gardens as decorations. They are rightfully described by the following lines "Beautiful flowers do not understand people feelings and may invite trouble, only silent stones are best for bringing peace and harmony."

Piling Up Rockeries and Creating Sceneries

As the Chinese saying goes "A garden without stones is not a perfect garden", stones are regarded as one of the fundamental elements in Chinese classical gardens. Due to their capacity to create sceneries beyond the physical landscape and their potential to inspire imagination, stones play a unique and

点。所以古典园林中最重要的境就是"山景"。堆石为山，由人工造出深幽、峭拔、灵秀的感观；又常以水辅助，营构缩微的自然山水。园林中还常选用少量的山石进行巧妙的点缀，布置成景。其位置常选在庭院中、池塘边、路边、墙角等处，所用石材常为太湖石、昆山石、灵璧石、英石等，布局方式有特置、对置、散置、群置等。

indispensable role in garden architecture. The clever deployment and display of stones represent a special feature of traditional Chinese gardens and create a unique charming garden environment.

Stones are good materials for piling up rockeries which are backbones in a classical Chinese garden and the key component of a garden building project. Therefore, the most important element in a traditional Chinese garden is the "rocks". Miniature natural landscapes are created by piling up rocks and stones to create steep and exquisite features. Lakes and streams are used to supplement the rocks. The gardens are often skillfully decorated with rocks and stones to create local scenic spots. These spots are usually located in the center of the courtyard, beside the pond, on roadsides or corners. The most commonly used types of stones in a classical Chinese garden include the Taihu Stone, Kunshan Stone, Lingbi Stone and Ying Stone. Layout methods include special placement, symmetrical or paired arrangements, and scattered or group arrangements.

• 苏州网师园内特置的太湖石
Taihu Stone Specially Placed in the Master-of-nets Garden, Suzhou

- 苏州留园中群置的假山石
  Group Rockeries in the Lingering Garden, Suzhou

- 北京颐和园中对置的山石
  Symmetrically Placed Pair Rocks in the Summer Palace, Beijing

● 苏州狮子林中的假山石

狮子林为苏州四大名园之一，园内有国内尚存最大的古代假山群。湖石假山玲珑众多、出神入化，形似狮子起舞，被誉为"假山王国"。

Rockeries in the Lion Grove Garden, Suzhou

As one of the four famous gardens in Suzhou, the Lion Grove Garden maintains the largest traditional rock groups in China. The rocks are in diverse shapes, vivid and exquisite, like dancing lions. The garden was hence honored the "Kingdom of Rockeries".

● 苏州沧浪亭的假山

Rockeries in the Surging Wave Pavilion, Suzhou

室内陈设

"斋无石不雅",古人常以美石装点于家中,为房中增添几分雅致,若有合适的山水书画相衬,更

- 《岁朝清供图》吴昌硕（清）
  "岁朝"指农历正月初一,历代文人画家都喜欢在这一天绘制《岁朝清供图》,作为新年伊始的首幅作品。这幅图以瓶中梅花、水仙、兰草和一尊奇石组合而成,物品高下安排错落有致。

New Year's Painting by Wu Changshuo (Qing Dynasty, 1616-1911)
In ancient times, painters and calligraphers liked to work on their first painting or calligraphy on New Year's day to welcome the New Year. This painting depicts plum blossoms, daffodils, orchids and a piece of wonder stone. All the items are artistically arranged.

Indoor Display

"A room without stones is not elegant". Chinese scholars in ancient times always decorated their rooms with beautiful stones, adding elegance to their plain living environment. If the stones were accompanied by landscape painting and calligraphy, it would be even better. A couple of thought provoking landscape

- 苏州拙政园卅六鸳鸯馆内的案头供石

A special tribute stone in the 36 Pairs of Mandarin Ducks' Pavilion of the Humble Administrator's Garden, Suzhou

是锦上添花。一两幅境界深远的山水画、俊逸不凡的书法条幅,能够与石形成一个完美的整体氛围,令居室雅意大增。

奇石本身不经修饰的特性,象征着与自然的和谐相处之道。一直以来,房间内摆设的观赏石成为文人所喜爱的雅玩。

paintings or graceful and outstanding calligraphy scrolls with stones can create a harmonious atmosphere, tinting the rooms with elegance.

Unpolished stones symbolize harmony with nature. Wonder stones placed in rooms have become items of appreciation and enjoyment for literati and scholars in China for a longtime.

● 苏州留园的奇石盆景
Bonsai landscape in the Lingering Garden, Suzhou

中国石之美
The Beauty of China Stones

　　从石的天然特征来说,根据不同石种在形、色、质、纹四方面的特长,形成了不同的观赏聚焦点,依此可分为以形状美取胜的造型石,以纹理美取胜的纹理石,形状规律、晶莹剔透的矿物晶体,以及记录了地球沧桑变化的古生物化石等四大类。

Based on the different strengths of stones in terms of four natural features of shape, color, texture and grain, different stones distinguish themselves with their own appreciation highlights. Accordingly, wonder stones can be divided according to their best features: modeling stones for their shapes, striated stones for their beautiful patterns, mineral crystals for their regular shapes and transparent crystals, ancient biological fossils for their records on earth crust movements.

> 造型石

造型石是奇石中最常见的一种类型。它主要是在溶蚀作用、水蚀作用、风蚀作用、火山作用、构造运动等各种地质作用下，由岩石和矿物等形成的石体。

造型石的大小不一，大的可高达数米，小的二三十厘米，甚至更小。大型的造型石多置于园林或花园、庭园，中等大小的造型石常置于厅堂内，而小型的造型石往往置于书斋或案头。

灵璧石

灵璧石是中国"四大奇石"之一，主要产于安徽省灵璧县的浮磐山，与之交界的徐州、泗县也有分布，古时有"泗水之滨多美石"之

> Modeling Stones

Modeling stones or shaping stones are the most common type of wonder stones. The stone body is composed of rocks and minerals through geological processes such as dissolution, erosion, weathering, volcanism and tectonic movements.

Modeling stones vary in size tremendously. While the big ones can be as high as several meters, the small ones are only twenty centimeters and even smaller. The big-size stones are mostly placed in gardens or courtyards while the medium-sized ones are laid in halls and rooms and the small ones are put in studies or on desks.

Lingbi Stones

Lingbi Stones is one of the "Four Key Wonder Stones" in China. Lingbi Stone is found mainly in Fupan Mountain of

说。灵璧石由颗粒大小均匀的微粒方解石组成，含有多种金属矿物质及有机物质。在大自然的鬼斧神工之下，灵璧石多为立体造型石，肖形状景，写意传神，气韵生动，具备了"瘦、皱、漏、透"等形态美的特点，是天设地造、美妙绝伦的天然艺术品，自古以来就受到历代文人雅客的推崇。

灵璧石的开采极早，商周时期，人们就发现了灵璧石叩之有声

Lingbi County, Anhui Province. The stone is also found in the neighboring Xuzhou City and Si County. As the old saying goes that "there are many beautiful stones along the Sishui River". Lingbi Stone is composed of even-size calcite particulates, containing various metallic minerals and organics. Through the wonderful artisanship of nature, most Lingbi Stones are transformed into three dimensional modeling stones with beautiful shapes,

• 苏州拙政园冷泉亭
Cold Spring Pavilion in the Humble Administrator's Garden, Suzhou

• 冷泉亭内的灵璧石
Lingbi Stone in Cold Spring Pavilion

的特点，用以制作石磬，所以又称"磬石"。南唐后主李煜爱"灵璧研山"，这是古人赏玩灵璧石最早的记载。宋代追求清淡脱俗的艺术表达，因此灵璧石这一自然天成的艺术品受到广泛的喜爱，苏轼、范成大所藏的"小蓬莱""小峨眉"皆为灵璧石珍品。宋徽宗曾得到一块玲珑秀润的

vivid images and lively charm, meeting the requirements of "slender, wrinkle, hollow and transparent". Lingbi Stones are perfect and splendid natural works of art that have been admired and praised by generations of Chinese scholars and literati.

Lingbi Stones were exploited since ancient times. During the Shang and Zhou dynasties, people found that the stone could produce a sound when knocked or beaten. So, they used the stone to make chimes, a musical instrument. Hence Lingbi Stone also got the name "Chime Stone". Li Yu, king of the Southern Tang Dynasty, liked to "study the mountains because of his love for Lingbi Stones". This is the earliest record about ancient people's appreciation for Lingbi Stones. People in the Song Dynasty pursued artistic expressions that were simple and elegant. Lingbi Stones, as natural works of art became very popular during the Song Dynasty. The bonsai "Miniature Penglai" collected by Su Shi and the bonsai "Mount Emei" collected by Fan Chengda are all excellent and precious representatives of Lingbi Stones. Emperor Huizong of the Song Dynasty once came to the possession of an exquisite piece of Lingbi Stone,

● 韵（灵璧石）
The Rhythm (Lingbi Stone)

"灵璧小峰"，十分喜爱，于石上御题"山高月小，水落石出"。南宋杜绾所著的《云林石谱》中汇载了当时有名的石品百十余种，并将灵璧石放在首位。宋代诗人方岩曾赞叹灵璧石："灵璧一石天下奇，声如青铜色如玉。"

明代收藏、观赏灵璧石之风持续盛行。明代文人文震亨在《长物志》中写道："石以灵璧为上，购之颇艰，大者尤不易得，高逾数尺者更属奇品。小者置几案间，色如漆声如玉者最佳。"到了清代，灵璧石身价更高，曾被乾隆帝誉为"天下第一石"。

灵璧石之美

质地：致密坚硬，因长期裸露于地表，久经暴晒和风、霜、雪、雨的冲刷，故而筋骨精练嶙峋，洞壑交错。

声音：叩之有悦耳的铜钟声，余音悠长，润人肺腑，故又称"八音石"，早在商周时代就用以制作石磬。今天仍有很多人将灵石之声视为驱邪纳福的吉祥之音。

纹理：灵璧石的皮表纹理斑驳，有裙褶、绉带、玉脉、赤线等

which he adored so much. So he wrote an inscription on this "Mini Lingbi Peak" reading "When the moon rises to the high peak, it looks so small; when the water recedes, the stones emerge." *Yulin Stone Collection*, a book compiled by Du Wan of the Southern Song Dynasty, recorded several dozens of famous stones at that time, of which, Lingbi Stone was at the top of the list. Fang Yan, a poet of the Song Dynasty once praised Lingbi Stone as "Lingbi Stone is the wonder of the world. It has the sound of bronze and the luster of jade."

The Ming Dynasty witnessed a lasting flourishing period for the collection and appreciation of Lingbi Stones. In his book *The Element of Garden Making*, Wen Zhenheng, a scholar of the Ming Dynasty, wrote that "Lingbi Stones are the best among stones. It is difficult to get them, especially large pieces. Stones that are dozens of meters high are rare and precious. Small Lingbi Stones can be placed on your study desk. Those with the luster of lacquer and sound of jade are the best." By the Qing Dynasty, Lingbi Stones became more valuable. It was praised by Emperor Qianlong as the "Number One Stone in the World".

不同形态，呈平、凸、点、线和面分布，纹理之间常常交错缠结。

韵味：灵璧石集质、色、形、趣、韵、气于一身，每块石头虽不能全面符合这些条件，但有韵、有质便是好石。奇石的韵味抽象，需用心感悟。

Beauty of Lingbi Stones

Texture: hard with high density. Because it is always exposed to air, sunshine, wind, frost, snow and rain, Lingbi Stone has a refined and rugged structure with interlocked caves and furrows.

Sound: when knocked, it produces dulcet sounds like that of a bronze bell, lingering and heart-caressing. So it is also called the "Music Stone". It was used to make chimes as early as during the Shang and Zhou dynasties. Today, many people still regard the sound of Lingbi Stone as a sound of auspiciousness with the power to expel evils and bring about happiness.

Grain: mottled grains on the stone surface taking varied forms such as folds on skirts, wrinkled belts, jade veins or red lines. They are distributed as flat, convex, dots and line patterns and are often intertwined.

Charm: Lingbi Stones embody all the excellent features of texture, color, shape, taste, rhythm and spirit in one. Although not every piece of stone can possess all these qualities, a stone impregnated with charm and has a quality texture is a good piece. The charm of a wonder stone is very abstract, can only be perceived with a willing heart.

• 灵璧石
Lingbi Stone

灵璧石的主要种类
Main Types of Lingbi Stones

品种 Type	特点 Characteristic
黑灵璧 Dark Lingbi	通身漆黑，质地坚韧致密，音质清脆，其色是因含有机物腐殖质而成。 Pitch dark color and hard texture with high density. Sound is clear and crisp. The color was formed by organic humus.
灰灵璧 Gray Lingbi	通体灰黑，外形怪异，以天然呈现动物形态而著称。 Grayish black color, grotesque shape. Famous for their natural animal-like shapes.
白灵璧 White Lingbi	纯净的方解石晶体，质地细腻，有冰肌玉骨之韵。 Pure calcite crystals, with refined texture and elegant charm.
红灵璧 Red Lingbi	主要矿物是方解石，因含氧化铁呈紫红、砖红等色。 Main mineral ingredient is calcite. Showing red and voilet colors because of the ferric oxide component.
五彩灵璧 Multi-Color Lingbi	质地细腻，由棕色、棕褐、土黄、赭黄等不同颜色的层次纹带缠绕组成，以色彩取胜，基本上没有动物的形状，叩击声音较为沉闷。 Fine and smooth texture and inter-twined layers of brown, yellow and deep yellow. It is famous for its colors, with basically no animal shapes. Its sound is dull and ponderous.
灵璧龟纹石 Turtle Moire Lingbi Stone	音、形、质与黑灵璧相似，其跌宕起伏的石表凸现着纵横密布的龟裂纹。 Similar sound, shape and texture with that of dack Lingbi Stone. Its uneven surface is full of turtle moire patterns.
灵璧叠层石 Multi-layer Lingbi Stone	石表纹理中有明显裸露藻类植物的痕迹，其中螺旋纹状圈层构造为蓝绿藻类痕。 Surface grains have obvious trails of nude alga. Among them, the spiraled layer structures are the trails of blue and green alga.

- 龟（五彩灵璧）
  Turtle (Multi-colored Lingbi Stone)

- 弓（白灵璧）
  Bow (White Lingbi Stone)

- 麒麟献瑞（红灵璧）
  Kylin Presenting Auspiciousness (Red Lingbi Stone)

云林石谱

《云林石谱》大约成书于南宋绍兴三年（1133），是中国古代体例最完整、内容最丰富的一部观赏石专著。作者杜绾，字季阳，浙江山阴（今浙江省绍兴市）人，平生好石。宋代社会风气重文轻武，上自皇帝，下至臣民，玩石藏石者众多。《云林石谱》对宋人赏石的理论进行提炼、总结和发展，体现了宋代文人赏石观的精髓。全书约14000余字，涉及名石共116种。作者详细考察了这些名石的产地，还详细记述了其采集方法、形状、颜色、质地优劣、敲击声音、坚硬程度、纹理、光泽、晶形、透明度、吸湿性、用途等方面的特点。

Yunlin Stone Collection

Yunlin Stone Collection is a book compiled in the third year of the Shaoxing period of the Southern Song Dynasty (1133 A.D.). It is a monograph on stone appreciation in ancient China with the most complete format and richest content. The author Du Wan, courtesy name Du Jiyang, was born in Shanyin City (current day Shaoxing City), Zhejiang Province. Du Wan was a stone lover all his life. Thanks to the social trend during the Song Dynasty which attached more importance to cultural over military achievements, collecting and appreciating wonder stones became a popular hobby among emperors, officials and citizens. By refining, summarizing and extending theories of the Song people on stone appreciation, the book *Yunlin Stone Collection* represents the salience of Song scholars' views on stone appreciation. With 14,000 characters, the book introduced 116 types of famous wonder stones. Having visited the production sites of these famous stones, the author was able to describe detailed features of these stones including collection methods, shapes, colors, texture, sound, hardness grade, grains, luster, crystal forms, transparency, hygroscopic levels and purposes, etc.

太湖石

太湖石，也称"洞庭石"，主要产于江苏省太湖地区的禹期山、鼋山、洞庭山。太湖石一般体量较大，最高可达三五丈，以造型取胜，多玲珑剔透、重峦叠嶂之姿。

Taihu Stones

Taihu Stones, also known as "Dongting Stones" are found mainly in Yuqi Mountain, Yuan Mountain and Dongting Mountain in the Taihu Lake Region, Jiangsu Province. Taihu stones are usually huge in size, with the largest ranging between 10-15 meters high. The stones are famous for their modeling functions because of their exquisite shapes and overlapping mountain peak postures. Most of the stones are in pure gray color, few are pure white or black. The pure color stones are of better quality as they complement better with lights.

Due to their voluminous size, Taihu Stones have always been used in traditional imperial gardens in China to create scenic spots. In order to build the "Genyue Garden", Emperor Huizong of the Northern Song Dynasty ordered local governments to collect precious plants and wonder stones from all over the country. Most of the stones collected were Taihu Stones. When constructing the "Daning Palace" after the demise of the Northern Song Dynasty, Emperor Shizong of the Jin Dynasty ordered to ship many Taihu Stones left in the "Genyue Garden" to Beijing and had them piled up on the Jade Islet in Beihai

• 云飞缈峰（太湖石）
Peak of Flying Clouds (Taihu Stone)

太湖石的色彩多为灰色，少见白色、黑色。一般无杂色，色泽纯净者最佳，色感与光感相得益彰者为好。

太湖石因尺寸高大，是中国古代皇家园林常用的布景主要石材，北宋时期宋徽宗为营造御苑艮岳，

Park. In the ensuing Ming and Qing dynasties, many Taihu Stones were moved from the Jade Islet to the Imperial Garden and the garden attached to the Palace of Tranquil Longevity, to create scenic spots there.

- 北京故宫御花园里的太湖石

Taihu Stones in the Imperial Garden of the Forbidden City, Beijing

- 冠云峰

冠云峰，现存苏州留园，该石"清、奇、顽、拙、瘦、皱、漏、透"八个字占全，尤其一个"皱"字，为他石所不及。

Cloud-capped Peak

The Cloud-capped Peak is currently kept in the Lingering Garden, Suzhou City. This piece of stone meets all the criteria of "clear, wonder, naughty, clumsy, hollow, slender, wrinkled and transparent". Its "wrinkles" cannot be matched by any other stones.

- 玉玲珑

玉玲珑，现存于上海豫园。该石四面八方洞洞通窍，如一孔清水，则孔孔出水，如焚香于一孔，则上下孔孔冒烟，可见其奇巧无比。该石为江南园林三大名石之一，艮岳旧物。

Exquisite Jade Stone

The Exquisite Jade Stone is now kept in the Yuyuan Garden, Shanghai. The stone has many inter-connected caves and furrows on its body. If you fill one cave with water, water will spill out from all other caves. If you burn incense in one cave, smoke will emit from all other inter-connected caves. This stone is one of the three famous stones in gardens in southern China. It belonged to the Genyue Garden originally.

- 瑞云峰

瑞云峰，现存江苏省苏州市第十中学校内，石形若半月形，多孔，玲珑多姿，是江南园林三大名石之一。

Auspicious Cloud Peak

Now kept in the Tenth Middle School of Suzhou City, the Auspicious Cloud Peak is a half-moon shape, with exquisite caves and furrows, and in a graceful posture. It is one of the three famous stones in gardens in southern China.

在全国各地搜罗奇花异石，其中以太湖石居多。北宋亡国后金世宗兴建大宁宫时，将艮岳残留的大量太湖石运往北京，堆砌在北海琼华岛上。明清两代，为点缀皇宫御花园和宁寿宫乾隆花园，曾从琼华岛取走大量太湖石。

太湖石形状可谓千姿百态，或玲珑剔透、灵秀飘逸，或浑穆古朴、凝重深沉、超凡脱俗，正迎合了古人赏石重形不重色的审美习惯，暗含了东方艺术"以意为象"的本质。同时太湖石那玲珑婀娜的姿态所透出的美的意蕴和境界，与古代文人雅士对平和、稳定、娴静、自律的心境和天命精神相关。

北太湖石

在北京市房山区周口店地区，出产一种石灰岩，形状大体和南方太湖石相似，具有太湖石的涡、沟、环、洞的变化，有"北太湖石"之称。同南方太湖石比，北太湖石体重更大，叩之无共鸣声，石表多密集的小孔穴而少有大洞，体态嶙峋，质地坚硬。北太湖石因大部分埋在土层里，开采后多呈土红色、橘红色、土

Taihu Stones are diversified in shape and appearance. Some are exquisite and elegant, others are simple but imposing, extraordinary and refined. These features meet precisely Chinese scholars' aesthetical approach of appreciating the shape rather than the color in ancient times. It also gives implications to the essence of oriental art, which is taking the "meaning first". Meanwhile, the appealing charm and beauty impregnated in the exquisite and graceful shapes of Taihu Stones echoed the aspiration of traditional Chinese scholars and literati for peace, stability, demure and self-discipline, as well as their belief in social responsibilities.

Northern Taihu Stones

Zhoukoudian area of Fangshan District, Beijing City, yields a kind of limestone, whose shape is similar to that of the Taihu Stones in southern China. As there are variations such as caves, furrows, rings and holes on these stones, they are also known as "Northern Taihu Stones". These stones are generally bigger than Taihu Stones from southern China. When knocked, they do not produce resonant sounds. Although many small holes and caves are found on the body of

- 北京颐和园乐寿堂前的"青芝岫"

  青芝岫置于颐和园乐寿堂前院内，石质是北太湖石，产自北京房山，长8米、宽2米、高4米，重约二十几吨，形似灵芝，是中国最大的园林观赏石。

  ### Qing Zhi Xiu in the Happiness and Longevity Hall of the Summer Palace, Beijing

  "Green Ganoderma Lucidum (Qing Zhi Stone)" is a famous piece of Northern Taihu Stone placed in front of the Hall of Happiness and Longevity in the Summer Palace in Beijing. A piece of Northern Taihu Stone in substance, it was produced in Fangshan District of Beijing City. This 8 meters long, 2 meters wide and 4 meters high rock takes the form of a gigantic ganoderma lucidum. Weighing more than 20 tons, it is the biggest piece of appreciation garden rock in China.

黄色，日久后表面带些灰黑色，质地不如南方太湖石脆，有一定韧性。其浑厚雄壮的外观，与南方太湖石的轻巧、清秀、玲珑有明显区别。

Northern Taihu Stones, large caves are rarely seen. The stone surface is rugged and revealing and the texture is hard. Because most parts of Northern Taihu Stones are buried underground, they usually show brownish-red, salmon and brownish-yellow colors when excavated. In time they turn grayish-black. They are not as crisp as Taihu Stones from southern China and tougher in general. The voluminous and magnificent appearance of Northern Taihu Stones is quite different from the slender and exquisite appearance of Taihu Stones from southern China.

• 北京中山公园青云片

青云片石高3米，长3.2米，周长7米。石色发青，玲珑剔透，姿态优美。青云片与青芝岫原都是爱石如命的明代书画家米万钟的遗物，当年开采出来后，由于米万钟家财耗尽，在运输途中被弃于郊野。一百多年后，乾隆皇帝发现了二石，将大石青芝岫运至颐和园，小石青云片运至圆明园时赏斋。1925年，青云片石从圆明园废墟中被清理出来，移置中山公园，石上尚存乾隆帝亲题的"青云片"三字和八首诗。

Blue Cloud Slab (*Qingyun Pian*) in Zhongshan Park, Beijing

Blue Cloud Slab is a rock stone of 3 meters high, 3.2 meters long and 7 meters in perimeter. It's bluish in color, exquisite and transparent in shape and graceful in posture. Blue Cloud Slab and Green Ganoderma Lucidum were relics left by Mi Wanzhong, a painter and calligrapher of the Ming Dynasty, who was a passionate stone lover. After the stones were excavated, Mi's family went bankrupt. So the stones had to be left in the wilderness on the way to his hometown. More than a century later, Emperor Qianlong discovered these two huge rocks and ordered them to be shipped to Beijing. The larger piece the Green Ganoderma Lucidum was placed in the Summer Palace and the smaller piece the Blue Cloud Slab in the appreciation Hall of the Old Summer Palace. In 1925, the Blue Cloud Slab was removed from the remains of the destroyed Old Summer Palace to Zhongshan Park. On the stone, inscriptions of the Blue Cloud Slab and eight poems by Emperor Qianlong are still visible.

英石

英石又名"英德石",因产于广东省英德市英德山一带而得名。英石与灵璧石同属沉积岩中的石灰岩,主要成分是方解石,但硬度不及灵璧石。英石色泽有淡青、灰黑、浅绿、黝黑、白色等数种,其中以黑者为贵。英石多为中小形块,如锋刃矗立,似层峦叠嶂,洞穴幽深,宛转相连,表面布满密深褶皱,尽显奇异。大的英石可砌积成庭园山景,小的可制作成山水盆景置于案几,极具观赏和收藏价值。

Ying Stones

Ying Stones or Yingde Stones are named after their production site in Yingde Mountain, Yingde City, Guangdong Province. Same as Lingbi Stone, *Ying stone* is limestone in sedimentary rocks. Although Yingde Stone's main ingredient is also calcite, it is not as hard as Lingbi Stone. Ying Stone has a number of colors, such as light gray, gray, light green, pure black and white. Among them, black is the most precious color. Ying Stones are usually of medium to small size. They look like erecting blades, overlapping mountain peaks and steep valleys. Their surface is covered with deep wrinkles, showing their singularity and mystery. Large size Ying Stones could be used for building rockeries in gardens while small ones for making bonsai landscapes to be placed on desks for decoration. Both have good values for appreciation and collection.

Ying Stones have been exploited for about one thousand years. There were records on mining of the Ying Stones during the Five dynasties. In the Song Dynasty, Ying Stone became famous appreciation stones on study desks. Those excellent ones with many wrinkles

● 龙马拂波(英石)
Dragon-horse Brushing the Waves (Ying Stone)

英石的开采已有千年的历史，五代时就有采石的记载。到了宋代，英石更是成了著名的案头赏石，以峰峦起伏、嵌空穿眼者为上品，制成山水盆景。宋人还常取具山峰形状的英石用以置笔，称为"笔峰"，这在宋人赵希鹄的《洞天清录》、杜绾的《云林石谱》中都有记载。陆游的《老学庵笔记》中也曾描述英石"佳者温润苍翠，叩之声如金玉"。明代盛行盆景，明代造园家计成的《园冶》记载英

and furrows were made into landscape bonsais. People in the Song Dynasty often used hill-shaped Ying Stones to make brush holders, which they appropriately named "Brush Peaks". This practice was recorded in books such as *Examination and Identification of Ancient Relics* and *Yunlin Stone Collection* as compiled by Zhao Xihu and Du Wan during the Song Dynasty respectively. In his book *Notes of an Old Scholar*, Lu You described that "Excellent Ying Stones look exquisite, smooth and green. They can produce

- 苏州网师园万卷堂内陈设的英石
  Ying Stones at the Hall of Ten Thousand Books of Master-of-nets Garden, Suzhou

石"亦可点盆,亦可掇小景"。清代英石的使用、买卖、收藏更加普及,这些英石配以精致的汉白玉雕座,优雅高绝,放置在皇宫中尽显皇家的非凡气度。还有部分英石出口到西欧国家营造园林或缀景。

昆石

昆石又称"昆山石",因产于江苏省昆山市昆山玉峰而得名。昆石与太湖石、雨花石等并称于世,为江南名石之一。其高度一般仅有尺许,以雪白晶莹、窍孔遍体、玲珑剔透为主要特征,故又称为"玲珑石"。按其形态特征分别命名为鸡骨峰、杨梅峰、胡桃峰、荔枝峰、海蜇峰等,其中以鸡骨峰石最为名贵。鸡骨峰石由薄如鸡骨的石片纵横交错组成,给人以坚韧刚劲的感觉;而胡桃峰石表面皱纹遍布,块状突兀,晶莹可爱。

昆石的成分主要是白云岩(碳酸钙和碳酸镁),是距今5亿年前的寒武纪海洋环境的产物。在一块昆山石上时常可以看到后期硅质溶液沿白云岩裂隙、空洞贯入形成的水晶晶簇,似雪花点缀,晶莹可爱,

clear and crisp sound when knocked, like gold and jade." Landscape bonsais were very popular during the Ming Dynasty. In his book *Garden Building*, Ji Cheng, a famous architect of the Ming Dynasty, stated that "Ying Stones could be used to make landscape bonsais or make miniature sceneries." The use, transaction and collection of Ying Stones became more popular during the Qing Dynasty. Placed in imperial palaces on exquisitely carved white marble pedestals, these Ying Stones magnify the extraordinary elegance and dignity of the imperial family. Some of the Ying Stones were exported to Western European countries to build gardens or to create scenic spots.

Kun Stones

Kun stones, also known as Kunshan stones are named after their production area at the Jade Peak of the Kunshan Mountain, Kunshan City, Jiangsu Province. Similar to the Taihu Stones and the Rain-flower Stones, Kun Stones are reputed as one of the famous stones in southern China. Because of the small size of Kun Stones (generally about a third of a meter high), plus their white color, and dotted body with caves and hollows all over, it is nicknamed the

瑰丽奇异。昆石的开采历史悠久，宋代《云林石谱》中就有记载：将山洞中的白云岩毛坯采下后，先在太阳光下暴晒五六天，使其外表的红泥发硬剥落；再用碱水反复冲刷，并仔细剔除石孔内的泥屑石粒；然后，用一定浓度的草酸洗去石上的黄渍并晒干。这样，昆石便成为洁白如雪、晶莹似玉的观赏石精品了。

"Exquisite Stone". Kun Stones are further divided into sub-categories according to their different shapes, including chicken bone peak, plum peak, nut peak, lychee peak, and jellyfish peak, etc. Among them, chicken bone peak is the most precious type, because it is composed of crisscrossed stone slabs that are as fine as chicken bones. It impresses people with a sense of perseverance and power. The nut peak is covered with all kinds of wrinkles, convex and concaves, very charming and pretty.

The main ingredient of Kun Stone is dolomite (calcium carbonate and magnesium carbonate), a product of the marine environment during the Cambrian Period 500 million years ago. On a Kun Stone, one could always see crystal clusters that were formed by silicon liquid dropped through dolomite cracks and holes during the later phase of the formulation. These crystal clusters look like bright and pretty snowflakes. Kun Stone enjoys a long history of exploitation. The book *Yunlin Stone Collection* published during the Song Dynasty described the mining process of Kun Stones as follows, "When dolomite rough stone is taken from the mountain caves, it is first exposed to strong

• 乱云飞渡（昆石）
Flying Clouds (Kun Stone)

• 脑海妙思（昆石胡桃峰）
Contemplating Wonders (Nut Peak of the Kun Stone)

sunshine for five to six days. When the red clay on the rough stone's surface is dried and falls off, it is then washed repeatedly with alkaline water, and the clay and particles in the wrinkles and holes are carefully removed. Then the yellow stains on the stone are washed off with oxalic acid of certain concentration. Finally, the dolomite rock is dried. After all these processes, a rough piece of Kun Stone becomes an excellent piece of appreciation stone which is as white as snow and as shining as jade.

博山文石

博山文石因产于山东省淄博市博山区而得名，又名"淄博文石"。博山古时属于青州府，博山

Boshan Aragonites

Boshan Aragonite is also known as Zibo Aragonite. The stone is named after its place of origin in the Boshan District of Zibo City, Shandong Province. Boshan was under the administration of Qingzhou City in ancient times. *Boshan Aragonite* is actually one branch of the "Qingzhou Stone", which was a famous appreciation stone family as recorded in the book *Yunlin Stone Collection*. Boshan Aragonite is a type of limestone with very fine texture. Most of the Boshan

• 龙盘雪崖（博山文石）
Coiling Dragon on Snow Cliff (Boshan Aragonite)

文石实际是《云林石谱》中记载的著名观赏石"青州石"的一种。博山文石属于石灰岩，石质细腻，一般呈灰白色、灰褐色、玄黑色等。在风化、雨水冲刷及地壳运动等作用下，山体岩石破碎后埋于地下，长期受地下水和酸性土壤的渗透溶蚀，形成了千姿百态的造型和大小不等的孔窍沟壑以及表面变化多端的立体纹理。这些纹理皱皱，或点线交错，或延伸平行，或弯曲旋转；有些石体还间有白色的方解石细脉，或红色的铁质线，形成了博山文石独特自然的风格。

Aragonites show pale gray, brownish gray or black colors. In the long process of weathering, rain brushing and crust movements, mountain rocks are crushed and buried underground. Having been infiltrated and eroded in acid soil and underground water for many years, grotesque shapes, holes and cracks were formed with exotic three-dimensional grains on the rock surface. These grains are wrinkled, crossed, paralleled or curled. Some have white veins of calcite between stone bodies. Others have very fine red iron lines. These characteristics are unique features of the Boshan Aragonite.

- 玉宇环楼（大化石）
Jade Tower (Dahua Stone)

- 云海腾龙（大化石）
Flying Dragon (Dahua Stone)

大化石

大化石是"大化彩玉石"的简称，产在广西大化县岩滩的红水河段，学名"岩滩彩玉石"，是近几年发现的新石种。

大化石以硅质岩石为主，石质坚硬致密，硬度很大，而且表面光洁、细腻、圆润，手感极好，玉化程度高的有如水晶般透明。大化

Dahua Stones

Dahua Stones, an abbreviation of "Dahua Color Jade Stones", were produced along the Hongshui River in Dahua County, Guangxi Zhuang Autonomous Region. Its scientific name is "Rock Beach Color Jade Stone", which is a new stone type found in recent years.

Dahua Stones are mainly composed of silicon rocks of high density. It's a hard stone with a smooth, sleek and lustrous surface. The highly petrified jade is crystal clear and transparent. Dahua Stones have orderly changing layers, natural colors and clear rhythmic grains. Its colors are both simple and beautiful, with golden yellow, grayish yellow, brownish red, dark brown, bronze,

• 风情万种（大化石）
Fascinating Charm (Dahua Stone)

• 金羊开泰（大化石）
Golden Sheep Heralds Peace and Prosperity (Dahua Stone)

石层理变化有序，色韵自然，纹理清晰而具有韵味；色彩明丽古朴，呈金黄、褐黄、棕红、深棕、古铜、翠绿、黄绿、灰绿、陶白等多种色泽，更有的一石数色，有的大红大紫，有的纯黑纯白，有的又蓝又绿，有的金黄赤橙，可谓多姿多彩，艳丽无比。

有人总结，大化石兼备了中国各类观赏石的优点，例如玛瑙石的坚硬润滑，寿山石的细腻光洁，灵璧石的花纹变化多样等，内涵非常丰富；同时大化石还具有其他观赏石比较少有的浮雕式的图案及奇特的外形。

different shades of green and pottery white. Some stones combine several colors, such as red and violet, black and white, blue and green, yellow and orange. Indeed they are dazzling and beautiful!

As summarized by some specialists, Dahua Stone combines the merits of various Chinese appreciation stones: hardness and smoothness of agates, fine texture and luster of Shoushan Stones and varied patterns and grains of Lingbi Stones. All these contribute to Dahua Stone's rich connotation. In addition, the relief patterns and unusual shapes of Dahua Stones are rarely seen in other appreciation stones.

红水河出奇石

红水河是珠江水系干流西江的上游河段，在贵州省和广西壮族自治区之间，因流经红色砂贝岩层致使水色红褐得名。河流全长659千米，流域面积3.3万平方千米。河流流经高原、低山和丘陵，沿途群峰夹谷，河床深邃，主要险滩有50余处。因红水河水流湍急，流量大，落差大，特别是洪水期泥沙俱下，河床中的石头经过骇浪不断地冲刷磨蚀，形成了凹凸不平、形态万千的怪礁奇石。一条河中能产出十多种高品质的奇石，这在全国乃至世界上都极其罕见，如大化石、摩尔石、梨皮石、来宾石、大湾石、彩陶石、天峨石、国画石等，均产自红水河的不同河段河床中。

Wonder Stones Found in Hongshui (Red Water) River

Hongshui River is the main stream of Xijiang River, which is the upper stream of the Pearl River System. Located between Guizhou Province and Guangxi Zhuang Autonomous Region,

Hongshui River got its name because of the brownish water color which is caused by the red sand shell rocks it flows over. With a total length of 659 kilometers and 33,000 square kilometers basin area, the river flows through plateaus, low mountains and hills, valleys and gorges, deep riverbeds and some 50 rapids and shoals. The rapid water, huge volume and large waterfall, especially mud and sand discharged during the flooding seasons continuously brushed and abraded stones buried in the riverbed. Gradually the stones became uneven, irregular and grotesque-looking reefs and strange stones. As many as ten varieties of high-quality wonder stones can be found in Hongshui River. This is extremely uncommon in China, even in the world. Stones such as Dahua Stone, Moore Stone, Pear Peel Stone, Laibin Stone, Great Bay Stone, Color Pottery Stone, Tian'e Stone and Chinese Painting Stone have all been found in different sections and riverbeds of Hongshui River.

• 广西红水河风光 (图片提供: FOTOE)
Scenery of Hongshui River in Guangxi

彩陶石

　　彩陶石产于广西红水河下游的合山市河里镇马安村，因此又被称为"马安石"。彩陶石的产地很狭窄，红水河流经此处时，受到长达几千米的暗礁阻挡，暗礁左侧则形成一条三百多米长的回水湾，彩陶石就产自这条回水湾中。长年累月的潮涨潮落，夹带着砂石的水流把回水湾中的石头表皮冲刷得非常光滑，同时加上光的作用和水的浸染，石头的表皮被染上了一层柔和

● 谦谦君子（彩陶石）
A Gentleman (Color Pottery Stone)

Color Pottery Stones

Color Pottery Stones are produced in Ma'an Village, Heli Town, Heshan City, which is located downstream of the Hongshui River in Guangxi. Therefore, Color Pottery Stones are also known as "Ma'an Stone". Color Pottery Stones are only produced along a narrow basin stripe where the Hongshui River flows by and a backwater bay of over 300 meters is formed beside a sunken reef of thousands meters long. With years of tide rise and tide ebb and brushing of sandy rapids, stones in the backwater bay became very smooth. Lights and water further render the stone surface a layer of gentle and soft hue, like that beautiful glaze painted on quality potteries in ancient times. It was because of this reason that local people proudly named it the "Color Pottery Stone".

　　Most Color Pottery Stones are in polygonal geometric form, with fine and smooth surfaces. The different mineral ingredients composing the stone render the stone various bright colors. Generally speaking, pale green, gray and black are the more common colors while green is considered superior to others. Natural transitional colors and well structured

的色彩，好像古代制作精良的陶器被涂上了一层美丽的釉色。所以，当地人自豪地称之为"彩陶石"。

彩陶石以多边形的几何形体居多，表面光滑细腻，各种矿物组成的颜色鲜亮，一般以豆绿、灰色、墨色较为常见，而以绿色为上乘。石上天然过渡色极佳，色泽条纹层次分明。

彩陶石有"彩釉"和"彩陶"之分。石肌似瓷器釉面的称"彩釉石"，似无釉陶面的称"彩陶石"。从颜色上分，又可分为纯色石与鸳鸯石，鸳鸯石是指双色石，而具有三色以上的又称"多色鸳鸯石"。

color grains of the stones are much admired.

Color Pottery Stones can be further divided into "Color Glaze" and "Color Pottery" stones. While the former has porcelain-like gleaming surfaces, the latter has unglazed pottery surfaces. In terms of colors, Color Pottery Stones can be divided into mono-colored and dual-colored ones. Stones with three or more colors are called "Multi-color Stones".

• 草堂春意（彩陶石）
Spring Coming to a Straw Hut (Color Pottery Stone)

• 湖山澄彩（彩陶石）
Beautiful Mountains on the Lake (Color Pottery Stone)

来宾石

来宾石因产于广西来宾市而得名，主要出自红水河中下游来宾河段的百艾滩。这一河段多险滩，水流湍急，沿河是熔岩地带，蕴藏着丰富的石种资源，千百万年惊涛骇浪对河床中的石头不断进行搬运、翻移，反复冲刷、磨蚀，从而形成了造型各异的奇石。

来宾奇石形、色、质、纹、声五大要素俱全。其形如鬼斧神工：状如人物者，惟妙惟肖；状如动物者，若奔若飞；状如景物者，似画似诗；更有如笔力刚劲、纹如

- 古韵（来宾石）
  The Classic Charm (Laibin Stone)

Laibin Stones

Laibin Stone got its name from its place of origin at Laibin City of Guangxi Zhuang Autonomous Region. They are mainly found in the Bai'ai Shoal of the Laibin River, which is the middle to lower reach of the Hongshui River. This section embraces many rapids and shoals where the current flows fast. The area along the river is of lava land with rich stone resources. Hundreds of millions of years of tempestuous waves and storms have been constantly moving, rolling, brushing and eroding the stones in the riverbed, resulting in the formation of many wonder stones with grotesque shapes.

Laibin Stones possess all the merits of a wonder stone: its shape, color, texture, grain and sound. With extraordinary shapes, some stones look like humans, others look like running animals or flying birds. Still, others look like beautiful paintings or calligraphy works with firm and powerful strokes. This kind of Laibin Stones are in simple and serene colors such as black and brown. Few have bright colors. The stone has a hard and fine texture, smooth surface and clear grains which can be

水墨的文字奇石。其色古朴沉稳，以黑色或古铜色最为常见，亦有色泽鲜艳的。来宾石石质坚硬，质地细腻，石肤细润光洁，纹理清晰，可分为细纹、线条纹、粗纹、层叠纹、云纹等多种，其声清脆如钟。

further divided into refined, lined, layered and cloud-patterned grains, etc. When knocked, the stone can produce a clear and crisp sound like that of a bell.

- 凤翼冲霄（来宾石）
Flying Phoenix in the Sky (Laibin Stone)

- 清风引岫（来宾石）
Breeze from the Groto (Laibin Stone)

如同大师雕塑的摩尔石

摩尔石是近年来广西红水河水冲石中崛起的新秀。2000年,在广西柳州市马鞍山的奇石市场上出现了一种质地柔韧、线条流畅、造型简练、极富动态美的水冲石。当时,人们的眼光集中于质地坚硬、色彩艳丽的硅质岩类石种;这种质地相对较软、色泽单一、缺乏温润的包浆、手感粗糙的奇石没有引起重视,被称为"磨刀石"。

2000年的年末和2001年的年初,英国现代雕塑大师亨利·摩尔的雕塑艺术展先后在北京、上海两地举行,摩尔创作的那些极富想象力的、超现实性的、带有抽象意味的雕塑作品引起不小的轰动。亨利·摩尔的雕塑线条柔和,体态夸张,开阖自如,十分大气。人们发现,"磨刀石"所具有的线条美、轮廓美、造型美的抽象表现形式,与摩尔的雕塑在线条圆滑流畅、块面柔和宛转、形体豪放夸张等方面有着惊人的相似。从此,"磨刀石"受到了人们的关注,也吸引了不少眼光独到的奇石收藏家,并顺理成章地赋予了它"摩尔石"的美名。以一个著名艺术大师的名字来为某一石种命名,这在奇石的历史上可谓绝无仅有的一例。

Moore Stones—Masterpiece of Nature

Moore Stone is a new star of the appreciation stone family that emerged in recent years from water-washed stones of the Hongshui River, Guangxi Zhuang Autonomous Region. It first appeared in the year 2000, on the stone market at Ma'anshan, Liuzhou City, Guangxi Zhuang Autonomous Region. This water-washed stone has a flexible texture, fluent lines, simple shapes and dynamic beauty. When it first appeared, most stone lovers were still focusing on silicon stones which had hard texture and flamboyant colors. Few paid attention to this new variety with a relatively soft texture, mono color, no smooth slurry and coarse surface. It was nick-named the "Grindstone".

In late 2000 and early 2001, the contemporary British sculpture master Henry Moore held exhibitions of his works in Beijing and Shanghai. Sensations were caused by such fully imaginative, surreal and abstract sculpture pieces created by the artist. Henry's sculptures had gentle lines, exaggerated body shapes, which were graceful and impressive. Local people suddenly realized that the beautiful lines, profiles and shapes of the "grindstones" are very similar to that of Moore's sculpture works, with their fluent lines, mild and soft blocks and surfaces and exaggerated gestures. From then on, "grindstones" captured the attention of many people including stone collectors. Thus, the name "Moore Stone" was openly bestowed on this "grindstone". It is the only case in the history of wonder stones where a stone is named after a famous artist.

• 思（摩尔石）
Contemplating (Moore Stone)

• 轿椅（摩尔石）
Sedan Chair (Moore Stone)

• 摩尔石
Moore Stone

乌江石

乌江，发源于贵州西部高原海拔2260米的乌蒙山脉东麓，在重庆市涪陵汇入长江，全长1050千米。乌江流域河谷深峻，水流湍急，是成就奇石的绝佳地域，也是中国著名的江河奇石产地之一。产于乌江水域的乌江石，多形成于数十米深的河底，由于石质坚硬，在水流湍急的河底历经千百万年的水冲沙

Wujiang Stones

Originated from the foot of the Wumeng Mountain with 2260 meter elevation, the 1050 kilometer long Wujiang River runs through the western Guizhou Plateau and joins the Yangzi River at Fuling, Chongqing City. The steep valleys and rapid currents are ideal cradles for wonder stones. Therefore, Wujiang River is one of the famous production sites for wonder stones in China. Most of the Wujiang Stones are found on

- 将军战盔（乌江石）
  The General's Helmet (Wujiang Stone)

- 舞动的北京（乌江石）
  Dancing Beijing (Wujiang Stone)

磨，水洗度、磨圆度极佳，石肤光洁莹滑，温润如玉。有些玉化程度高的乌江石，石肤可呈半透明状，给人以晶莹剔透之感。乌江石的颜色以黝青、牙黄、铁灰色较为常见，色调含蓄浑厚，朴拙而有典雅之风，古意盎然。而石上纹理所构成的图案，岭断云连，舒展不一，仿佛将自然的山水风景浓缩影印于石上；也有些乌江石上呈现出环状条带，玲珑宛转，刚柔并济，疏密相宜，恰到好处地将乌江石圆融的造型勾勒出来。

riverbeds that are several dozen meters deep. Because of the hard texture of the stone, after thousands of years of brushing, polishing and corroding, the stone surfaces became so smooth and sleek, like that of jade. Some of the better petrified Wujiang Stones have semi-transparent surfaces. Most Wujiang Stones are in dark green, light yellow and iron gray colors, displaying their simple but elegant beauty. Some patterns formed by grains look like natural landscapes. Others with exquisite rings and belts, harmonize skillfully with the roundish shape of the stones.

● 天际流（乌江石）
Flowing to the Horizon (Wujiang Stone)

● 乌江石
Wujiang Stone

盘江石

盘江石，产于贵州省境内的南北盘江及其支流河床中。盘江分为南北两条，均发源于云南马雄山，两江相拥穿行于贵州西南的崇山峻岭之间。南北盘江上下游地势海拔相差千余米，使江水形成很大的落差，同时中上游地段江面狭窄，山高谷深，江流十分湍急，为孕育盘江石营造了一个特殊的地质和地貌环境。

盘江石的石质绝大部分为石灰石，滚落江中的石头在碰撞解裂

Panjiang Stones

Panjiang Stones were found in riverbeds of the South and North Panjiang Rivers and their tributaries in Guizhou Province. Originating in the Maxiong area of Yunnan Province, both rivers run through high mountains and steep valleys of southwestern Guizhou. With the large elevation difference between the upper and lower reaches, a sharp waterfall of more than one thousand meters was formed. The upper and middle streams of the rivers are narrow and are flanked by steep mountains and deep valleys. The resulting rapid currents offer an ideal

● 盘江石
Panjiang Stone

● 天书遗卷（盘江石）
Book from Heaven (Panjiang Stone)

后，经过江水的不断溶蚀，表面形成各种形态的凹槽、孔穴，再经水冲沙磨，显得光洁温润，肌理丰富，形成了似山川景观、飞禽走兽、古堡碉楼、瓜果器皿等奇特多姿的造型。

潦河石

潦河位于江西省西北部，发源于九岭山脉，全长数百千米，迂回曲折，起伏跌宕。河流两岸各种不同成分的岩层受山洪或地壳运动的冲撞，有些岩石滚入河沟山谷，经过河水千百万年的冲刷、撞击、浸泡，才形成丰富多彩的潦河石。

- 古陶（盘江石）
Ancient Pottery (Panjiang Stone)

geological and terrain environment for the formation of Panjiang Stones.

Panjiang Stones are mainly composed of limestones. After rolling into the river, these stones were collided and corroded in the water. Various grooves and holes are formed on the surface. After being brushed and polished by water and sand, the stone surface became very smooth and sleek, forming various patterns resembling mountains and landscapes, birds and beasts, castles and towers, fruits and vessels, etc.

Liaohe Stones

Liaohe River flows over the northwestern part of Jiangxi Province. Originating in the Jiuling Mountain, this

- 潦河风光（潦河石）
Scenery of Liaohe River (Liaohe Stone)

从欣赏角度看，潦河石主要包括筋纹石、潦河蜡石、紫金石和潦河红几个品种。

筋纹石通常有两种以上不同硬度的石质组成，经过河水冲刷，表面形成了凹凸的纹理，加上底色与纹路之间的色泽差异，形成了多种多样的形状和图像。筋纹石的色彩

several hundred kilometer long river runs through steep mountains and valleys with many twists and turns. Due to mountain torrents or crust movements, many rocks with various ingredients on both sides of the river bank were flushed into the river. After hundreds of years of brushing, smashing, soaking and polishing, colorful Liaohe Stones were formed.

From the appreciation perspective, Liaohe Stones mainly entail several varieties: Grain and Vein Stones, Liaohe Alabasters, Purple Stones and Red Liaohe Stones.

Grain and Vein Stones are usually composed of two or more component substance of different hardness grades. After brushing by river torrents, these stones form concave and convex grains and veins on the surface. The different colors between the background and the grains form beautiful patterns and pictures. While the predominant color of the Grain and Vein Stones is black, grey, white and yellow dot the background. The size of the stones is usually in medium, with fine water erosion.

The main ingredient contained in the Liaohe Alabaster is silicon dioxide. With grease-like luster, good tenacity and transparency, the stone's color is

• 潦河石
Liaohe Stone

以黑色为主，夹有灰、白、黄等色，体量以中大型为主，水冲度极佳。

辽河蜡石的主要成分为二氧化硅，石表具有油脂光泽，有韧性，透明度好，颜色有金黄与橘黄相间、红色与白色相间、蛋黄与枇杷黄相间、紫色与白色相间等，显得流光溢彩，绚彩夺目，深受玩石者的喜爱。

紫金石全身紫色，有些石质中还掺有金色或其他颜色，在阳光照射下闪耀金光。而辽河红则以红色为主，有时掺有金色、白色、黑色，带有奇妙的纹理和艳丽的色彩。

intertwined between golden yellow and orange yellow, red and white, egg yolk and loquat yellow, purple and white. Indeed, Liaohe Alabasters have dazzling colors that are very appealing to stone collectors.

As suggested in the name, Purple Stones are basically purple in color. Some are dotted with gold or other colors that glitter under sunshine. While Red Liaohe Stones are predominantly red, they are sometimes dotted with other colors such as yellow, white or black, forming wonderful grains and patterns of flamboyant colors.

钟乳石

钟乳石多悬于石灰岩构成的溶洞内。石灰岩溶洞洞顶有很多裂隙，裂隙处常有水滴渗透。水分蒸发后，会留下石灰质的沉淀，经上万年或几十万年的时间，石灰质积累成了乳头，乳头外面又不断被石灰质包裹，越垂越长，形成了姿态万千的钟乳石。当洞顶上的水滴落到地面，石灰质也在地面上沉积起来，则出现了对着钟乳石向上生成的石笋。中国熔岩

Stalactites

Most Stalactites are found in limestone karst caves. On the ceiling of these caves, there are many cracks and crannies, where water drips all year long. When water evaporates, lime sediments stay. After tens and thousands of years, the lime piled into nipples wrapped with lime. When the nipples grow longer they become stalactites. When water drops on the floor, the lime accumulates and forms stalagmites. China has a well-developed lava landscape and rich stalactite

● 广西桂林银子岩溶洞
Yinzi Karst Cave in Guilin, Guangxi

地貌发育成熟，钟乳石资源丰富，广西、云南、贵州、四川、湖南、江苏等省区均有分布。

钟乳石生成于溶洞中，没有经历日晒和风化，所以形体保持得很完整，石质细腻，光泽剔透，形状奇特，具有很高的欣赏价值，其中又以雪白晶莹、表面泛有晶簇闪光者为佳。钟乳石色彩丰富，有乳白、浅红、淡黄、红褐等颜色，形成了五彩缤纷的图案。钟乳石的形态千奇百怪，有笋状、柱状、帘状、葡萄状，还有的像花朵、动物、人物，栩栩如生。不过其形体差别很大，小者盈寸，大者逾丈。

钟乳石用途广泛，在中国古代，钟乳石多为药用，不过钟乳石的装饰效果更为人们所看重。可将其配上底座，放置于客厅茶几上，以供观赏；也可植于陶盆中，组成山水盆景。近年来，由于人们对钟乳石的随

resources, which are mainly distributed in Guangxi, Yunnan, Guizhou, Sichuan, Hunan and Jiangsu, other provinces and regions.

Because stalactites are generated in Karst Caves free from sunshine and weathering, they maintain a perfect body, exquisite texture, clear luster and beautiful shapes and are of high appreciation value. In particular, those stalactites with white and crystal luster are the best. Stalactites have rich colors, including creamy, pink, light yellow and chocolate, forming diverse and vivid patterns. Stalactites also have various strange shapes, such as bamboo shoots, poles, curtains, grapes, even flowers, animals and humans. However, their sizes vary tremendously: the small ones are as small as one inch, and the big ones are as large as several meters.

Stalactites are widely used. In ancient times, stalactites were used mainly as a medicine ingredient. Nevertheless, people have always been

● 钟乳石
Innocent Love (Stalactite)

● 如意（钟乳石）
Ruyi (Stalactite)

意切割，造成了这一珍贵自然景观的破坏，所以目前禁止擅自开采。

结核石

结核石为沙质岩浆凝固而成的团块，含有多种金属元素，形成于距今2.5亿年至5.4亿年的寒武纪早期，形状以圆形和椭圆形居多。

结核石形体变化奇特，除了圆形以外，还有壶形、坛形、罐形、

more drawn by the ornamental effect of stalactites. They set stalactites on a stand and put them on tea-tables for decoration and appreciation. They also put stalactites in pots to make bonsai landscapes. In recent years, stalactite resources have suffered serious damage as a result of random cutting. So currently the exploitation of stalactites is prohibited without approval.

Nucleation Stones

Nucleation Stones are solidified gobbets of sand magma, containing many metallic elements. Nucleation Stones were formed during the early phase of the Cambrian Period some 250 million to 540 million years ago. Most of the stones are in round and oval shapes.

● 憨态可掬（结核石）
The Clumsy Charm (Nucleation Stone)

帽形、果形、砂锅形、铁饼形、车轮形、飞碟形、碗碟形、葫芦形、哑铃形、花生形、香炉形、三连体、多联体及动植物形体等。石体表面凹凸不平，表层散布着硫化铁的黄色晶体；晶体所构成的图案光芒四射，璀璨夺目。

　　当结核石形成之后，经过干燥或成岩过程，失水固结，体积收缩，会出现空腔。如果空腔中有可以自由活动的石核，摇动时便可作响，故结核石又被称作"响石"。结核石的石核有砂粒、土块、泥裂块、黏土、水晶簇、方解石晶簇和水等。石核的空腔越大，石核少而较大时，则响度越高。

Besides the round shape, Nucleation Stones have many rather incredible shapes, such as the shape of kettles, jars, cans, hats, fruits, terrines, discuses, wheels, frisbees, bowls, gourds, dumbbells, peanuts, incense burners, tri-pieces, multi-pieces, flora and fauna. The stone's surface is uneven, dotted with yellow crystals of iron sulfide, which form beautiful glaring patterns.

　　When Nucleation Stones were formed, they had to go through a drying or petrification process, when the stone was dehydrated and shrunk. If a cavity contains a free steinkern, it will produce a sound when it is shaken. That's why Nucleation Stones are also called the "Sounding Stones". Its kerns are composed of sand, clod, mud, clay, crystal cluster, calcite cluster and water. The bigger the cavity is and a fewer piece of larger size steinkerns it contains, the sound it produces will be louder when it is shaken.

• **结核石** (图片提供：FOTOE)
Nucleation Stone

风棱石

风棱石，又称"风蚀石""戈壁石""大漠石"，主要分布在气候干旱的荒漠地区，如中国内蒙古、青海、新疆等省区的戈壁荒漠，是风沙对裸露在地表的石块长期吹蚀和磨蚀形成的。风棱石以硬度大、均质的硅质岩为主，表面光滑，造型千姿百态、气势雄浑，有的像金字塔，有的似鹰、鸡，还有的呈蘑菇状、柱状、蜂窝状等。风

Wind Prismatic Stones

Wind Prismatic Stone is also known as "Wind-eroded Stone", "Gebi Stone" or "Grand Desert Stone". They are mainly distributed in arid deserts in Inner Mongolia, Qinghai, Xinjiang and other provinces and regions in China. These stones were formed by long-time weathering and erosion of wind and sand. Wind Prismatic Stones are mainly composed of silicon-type rocks with high hardness grade, even texture, smooth surface and various shapes, including pyramids, eagles, roosters, mushrooms, poles and bee hives. Based on the different stone textures, Wind Prismatic

● 风棱石
Wind Prismatic Stone

● 沙漠魂（风棱石）
Desert's Soul (Wind Prismatic Stone)

蚀石因质地的不同，又可分为玛瑙质、碧玉质、蛋白石质、石英质、长英质、水晶质、花岗石质等。进一步细分，玛瑙又可分为缠丝玛瑙、葡萄玛瑙、珍珠玛瑙和七彩玛瑙；碧玉又分为红碧玉、黄碧玉、绿碧玉等。

　　风棱石的纹理以各种花纹或不同色彩组成的动植物、人像、文字、风景图案等为特点，妙在似像非像，似是而非。由于风棱石生成年代久远，又处于人烟稀少的大漠戈壁，亿万年中无人扰访，故天时

Stones can be divided into the following categories including: agate, jade, opal, quartz, felsic, crystal and granite. Furthermore, agate can be divided into twined, grape, pearl and colorful types and jade can be further divided into red jade, yellow jade and green jade.

Grains on Wind Prismatic Stones constitute various patterns of animals, plants, portraits, characters and landscapes. The trick lies in the mysterious likeliness of the patterns and images. As Wind Prismatic Stones were formed in desolate deserts millions of years ago, they enjoyed the unique advantage of an enabling environment, favorable geological conditions and the divine artisanship of nature. Together

● 酋长（风棱石）
The Emirate (Wind Prismatic Stone)

● 宇宙（风棱石）
The Universe (Wind Prismatic Stone)

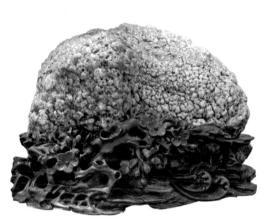

- 金甲鲮鲤（风棱石）
  Golden Carp (Wind Prismatic Stone)

地利人和的优势和鬼斧神工的自然力量使它形成了无奇不有、色彩丰富、千姿百态的形状。

千层石

　　千层石，因其石上横向纹理，层层横断，数以千计而得名。该石是一种久经风雨侵蚀风化的石灰质页岩，呈层叠状横纹，层次细密，中间夹有石砾层，可利用其横向纹理，表现"卧山式"造型，别具一格。颜色呈铁灰色、灰褐色，适宜制作海礁、海岛、沙漠风光，也可选取形状优美的层层独石，略加修整，配以基座，置于几案，别有一番情趣。

they created the stone's exotic shapes and extraordinary colors.

Thousand Layer Stone

Thousand Layer Stone got its name because of the many horizontal grains on its body. The stone is a kind of lime shale after many years of weathering and corrosion. Composed of many transverse lime layers with gravel layers sandwiched in between, the "Reclining Mountain" sculpture presents a unique style. The iron gray and gray brown color stones are more suitable for making reefs, island and desert landscapes. Some of the single pieces with beautiful shapes can be put on desks with a stand after slight trimming.

- 千层石
  Thousand Layer Stone

> 纹理石

纹理石，是指那些虽没有形成画面和形象，但是具有美丽纹理的观赏石。纹理石以其清晰、美丽的纹理或层理、裂理及平面图案为特色。纹理石的美在于它自然图案的神似，以及所表现出来的内涵和意境。纹理石大部分是卵石，自然界赋予它们内涵丰富的线条和图案。也有一些纹理石本身虽有很好的纹理，但在自然中未能直接显现出来，需要通过部分修饰和粗加工，才能使它的天然纹理得到充分的显示。

雨花石

雨花石，因原产于江苏南京雨花台一带而得名，其石质有玛瑙、蛋白石、水晶、玉髓等。相传在一千四百年前的梁代，有位云光法

> Striated Stones

Striated stones refer to wonder stones whose grains have not yet formed into patterns or images, but are featured by their distinct and beautiful grains, layers, cracks or flat patterns. The beauty of a striated stone lies in its resemblance with natural patterns and the connotation and implication they carry. Most striated stones are pebbles with rich lines and patterns bestowed by nature. Some striated stones have excellent grains, but they are not fully exposed in nature. Some processing or polishing work has to be done before these beautiful patterns can be revealed.

Rain-flower Stones

Rain-flower Stones are named after their original area of production in the Yuhuatai (Rain-flower Terrace) District, Nanjing City, Jiangsu Province. The

● 长河落日圆（雨花石）
Sunset over the Yangzi River (Rain-flower Stone)

师在南京南郊讲经说法，感动了上天，落花如雨，花雨落地为石，故称"雨花石"。法师讲经处更名"雨花台"，成语"天花乱坠"即由此传说而来。

经地质学家考证，雨花石形成于距今150万年至250万年之间。雨花石产生在火山熔岩的气孔中，或其他岩石的裂隙中，由二氧化硅的胶体溶液沿着空洞壁一层一层地沉淀固结而成。每层因含有不同的杂质而呈现不同的颜色和花纹，故而形成了不同品种。

雨花石一般直径在几厘米左右，多呈圆、椭圆、扁圆等形状，给人以娇小、玲珑、圆滑之感。雨花石的色素离子多样，色彩斑斓，以红、白、黄、黑、紫、褐、乳白、微黄为

stone's ingredients include agate, opal, crystal and chalcedony, etc. It was said that during the Liang Dynasty some 1,400 years ago, Master Yunguang preached in the suburbs of Nanjing City. His preaching moved the King in heaven so much, so flowers dropped from heaven like rainfall. As soon as the flowers touched the ground they turned into pebbles. Hence the name Rain-flower Stone came into being. The place where Master Yunguang preached was renamed the "Rain-flower Terrace". The Chinese idiom "flowers dropping from heaven" had its origin in this legend.

According to geologists, Rain-flower Stones were formed between 1.5 and 2.5 million years ago. Rain-flower Stones were generated in air holes of volcanic lava or crannies of other rocks where silicon dioxide colloid liquor was deposited and solidified into layers. Containing various impurities, each layer showed different colors and varied grains, forming different types of stones.

Rain-flower Stones are usually several centimeters in diameter and are round, oval and oblate in shape. They impress people with a sense of fragile exquisiteness and smoothness. Containing various color ions, the stones embrace

多，绿色、蓝色极少。雨花石的质地莹润光滑，透明者如滴珠，半透明者似凝脂，不透明者像洁瓷。雨花石的纹理多为圈状花纹，变化无穷，常可呈现各种山水、人物、花鸟等景象，引人遐想。

雨花石的分类常按矿物岩石种类划分，其中较为常见的有晶莹明澈的彩色玛瑙、蛋白石、燧石、水晶、化石等雨花石。

玛瑙类雨花石颜色艳丽，可分为红、蓝、绿、紫、酱斑、金黄、青黑等。其纹饰奇特，磨圆度高，晶莹可爱，属于雨花石中的上品，

diverse colors such as red, white, yellow, black, violet, brown and light yellow. There are few green or blue ones. The stones have a shiny and smooth texture. The transparent ones are like dewdrops and the semi-transparent ones are like coagulated grease, and the opaque ones are like porcelain. Grains on Rain-flower Stones usually have circular patterns with infinite variations. They often show images of mountains and rivers, humans, flowers and birds, inducing endless reveries and imagination.

Rain-flower Stones are usually divided into sub-categories according to the mineral rock types. The most available types include color agates, opals, flint, crystals and fossils, etc.

Agate-type Rain-flower Stones have various dazzling colors, such as red, blue, green, violet, brown patched, golden, black, etc. Because of their exotic patterns, smooth, round and shiny appearance, they are regarded as top-grade stones. As agate type Rain-flower

● 南京雨花台
Scenery of Yuhuatai District, Nanjing City

● 山涧清溪图（雨花石）
Clear Stream in the Valley (Rain-flower Stone)

因在水中更显剔透，又称"水石""细石"。

蛋白石类雨花石质地纯净，且具润感，有单色也有复色，色彩有失透状，观之如雾如烟，扑朔迷离。

燧石类雨花石，多呈褐、黄、黑色，少数因含色彩不同的条带构成的美丽花纹而成为珍品。

水晶类雨花石，除无色透明的外，还有烟晶、茶晶、黄晶、蔷薇水晶、紫晶、鬃晶、发晶等。能形成珍品的水晶类雨花石多为含有原生杂质、包裹体、结晶缺陷等构成的自然花纹。

化石类雨花石的石面上保留有一定观赏价值的动植物或珊瑚化石，有色彩形象者皆属精品。

Stones appear to be more transparent and exquisite in water, they are also called "Water Stones" or "Fine Stones".

Opal-type Rain-flower Stones have pure and sleek textures, single or complex colors. The opaque colors of the stones look like fog and clouds, rendering mysterious reveries.

Most of the Flint-type Rain-flower Stones are in brown, yellow and black colors. A few become precious because of the different color stripes and beautiful patterns.

Besides those achromatically transparent crystal-type Rain-flower Stones, there are many sub-types such as smoke-crystal, tea-crystal, yellow-crystal, rosebush-crystal, violet-crystal, mane-crystal and hair-crystal stones. Those precious crystal Rain-flower Stones are those with natural patterns formed by original impurities, inclusions or crystal defects.

Fossil-type Rain-flower Stones usually have fossils of flora and fauna or corals on the stone surface. Those with colored images are of superior grade.

In addition to the above-mentioned types, there are still other types of Rain-flower stones that are composed of jade rock, quartzite, tectonite, pebbles and

除了以上几类，还有碧玉岩、石英岩、构造岩、砾石岩等其他岩石和矿物组成的纹饰各异、色彩不同的极具观赏价值的雨花石。

other rocks and minerals. With their varied patterns and colors, they make quality Rain-flower Stones with high ornamental and appreciation values.

雨花石的保养

保持雨花石原有的纹理和色泽，必须知道相关的保养方法。一般的做法是用水将其养在干净的盛具内，最好是一日一换，保持水清石秀。

1.防干防裂

雨花石在自然环境下保持有一定的湿度，有些雨花石中含有水分子，若长期处于干燥条件下，会失去游离水分子，表面可能开裂。尤其是抛光石和蛋白石，干放时间过长就会开裂。保管时，最好能保持相对湿度，还要防止机械性碰撞，取放时要小心谨慎。

2.防酸防碱

雨花石的主要化学成分是二氧化硅，基本上属于中性。因此，所处的环境需要保持中性。偏酸或偏碱，酸碱中的离子就会与雨花石中的色素离子发生化学反应，使雨花石变质，或腐蚀，或变色。

3.防热防晒

如将雨花石加热或在强光下暴晒，随着温度的升高，石中的色素离子的性质会发生改变，使雨花石变质。因此，雨花石保存时要离开热源，避免在阳光下暴晒。

Maintenance of Rain-flower Stones

In order to maintain the original grain and luster of Rain-flower Stones, it is important to learn about some relevant maintenance methods. A common practice is to put the stones in a clean vessel or container and fill it with water. It is recommended that the water be renewed every day, so that the stones are kept in clean water.

1. Avoid drying and cracking

Rain-flower Stones should be kept in an environment with certain levels of humidity.

Some Rain-flower Stones contain water molecules. If they are exposed to dry conditions for a long time, their water molecules will be lost and their surface will crack. In particular, polished stones and opal stones will crack if they are kept in a dry environment for a long time. So it is recommended to keep these stones in a moist environment. It is also important to avoid mechanical collisions which might damage the stones.

2. Avoid acids and alkalis

The main chemical ingredient of Rain-flower Stones is silicon dioxide, which is basically a neutral substance, so Rain-flower Stones should be kept in a neutral environment. When the stone meets an acid or alkali environment, chemical reactions will take place between the ions in acid and alkali with the color ions of Rain-flower Stones, causing the degeneration, corrosion or color change of Rain-flower Stones.

3. Avoid heat and strong sunshine

If Rain-flower Stones are heated or exposed to strong sunshine, the color ions of the stones will change, and the stones will metamorphose. Therefore, we should keep Rain-flower Stones away from heat or strong sunshine.

• 雨花石
Rain-flower Stone

三峡石

三峡石是产于长江三峡地域内的宝石、矿石、古生物化石、珊瑚、图案卵石、象形石、文字石、纹理石、天然玛瑙石、供石等多种石类的总称。三峡石石源来自该区古老的前震旦系变质岩、沉积岩和前寒武纪侵入的花岗岩，主要分布在峡江两岸的溪流河谷或崇山峻岭中。

三峡石因构成不同，其表面虽有粗糙与光滑的区别，但却都活生生地展现了千姿百态的形状和异常丰富的天然纹理，并将人们带入浮

● 玫瑰傅影（三峡彩画石）
Graceful Shadows of Roses (Three-gorge Stone)

Three-gorge Stones

Three-gorge Stone is the generic name for stones found in the three-gorge area of the Yangzi River. The term encompasses many sub-categories of stones such as gems, mineral ores, ancient biological fossils, corals, patterned pebbles, shaping stones, character stones, striated stones, natural agates and offering stones. Three-gorge Stones are generated from Presinian metamorphic and sedimentary rocks and granites intruded during the Precambrian Period. Three-gorge Stones are mainly distributed in streams, valleys and mountains along the Yangzi River.

Inspite of the different surface textures of the stones with some being smooth and others rough, which was caused by the different ingredients of the stones, all Three-gorge Stones have incredible shapes and rich varieties of natural grains that trigger people's imagination. Three-gorge Stones embrace many varieties. Some opals and chalcedony-type stones are as small as a pigeon egg, as round and sleek as a wax pill and as beautiful as peacock feathers. The agate-type stones are as dazzling as silk brocades, as transparent as crystals and as gleaming as pearls.

想联翩的境地。三峡石种类繁多，有鸽蛋般大小、蜡丸般圆润、孔雀般美丽的蛋白石、石髓，又有彩缎般耀目、水晶般剔透、珍珠般晶莹的玛瑙等。

三峡石红、橙、黄、绿、青、蓝、紫七色俱全，其纹理、图案也很别致，纹路有凹有凸，人物形象神态各异，飞禽走兽栩栩如生。

从古到今，许多文人墨客都被三峡石的神奇倾倒。例如唐代诗人

Three-gorge Stones have many colors: red, orange, yellow, green, indigo, blue and violet. They are also featured by extraordinary grains and patterns, concave and convex lines, humans with different expressions and gestures and life-like birds and animals.

Since ancient times, many Chinese scholars and literati have been overwhelmed by the mystery of Three-gorge Stones. For example, a Tang Dynasty poet Bai Juyi described the wonderful shapes of Three-gorge Stones, stating that "The big ones look like swords, while the small ones look like teeth". Another famous poet of the Song Dynasty Ouyang Xiu described the picturesque beauty of the Three-gorge Stones with the verse reading:

• 白雪公主（三峡石）
Princess Snow White (Three-gorge Stone)

• 文字石（三峡石）
Character Stone (Three-gorge Stone)

白居易的诗句"大石如刀剑,小石如牙齿",描绘的就是三峡石的奇形;宋代诗人欧阳修的诗句"唯有山川为胜绝,寄人堪作画图夸",说的就是三峡石如图画般的美;明代诗人文安之的诗句"天精散落入地脉,龙章凤质何纷纭",赞的是三峡石的天生丽质;明代诗人严思浚所写的"峡山天下秀,余气散为石",感叹的则是三峡石的气韵。

九龙壁石

九龙壁石因产于福建漳州市华安县九龙江而得名,早在宋代就曾为

"Images of mountains and rivers on stones are absolutely beautiful like pictures". A Ming Dynasty poet Wen Anzhi praised the natural beauty of Three-gorge Stones as "Heavenly essence scattered on the earth, and transformed into beautiful stones". Another Ming Dynasty poet Yan Sijun exclaimed over the majestic grandeur and appealing charm of the Three-gorge Stones with the following verses "Mountains and gorges take up almost all the beauty and charm on earth, with the remaining charm dispersed and transformed into beautiful stones."

Jiulong Stones

Jiulong Stones are named after their production place at the Jiulong River, Hua'an County, Zhangzhou City, Fujian Province. The stones were used as tribute items to the imperial palace during the Song Dynasty. Jiulong River is bestowed with large waterfall gaps, rapid currents, clean and clear water with evergreen trees and foliage on both sides of the river. After many years of brushing, patting and polishing, the stones were gradually molded into many shapes. By virtue of their beautiful texture, shape, color, grain and charm, Jiulong Stones become

• 飞峰隐瀑(九龙壁石)
Waterfalls Hiding behind Mountain Peaks (Jiulong Stone)

皇宫贡石。九龙江落差大、水流急、水质好，江水长年累月清澈见底，两岸四季常青。九龙壁石历经漫长岁月，受急流的冲刷、拍击、磨洗，自然造就出千姿百态。九龙壁石以质美、形美、色美、纹美、意美，成为观赏石的新宠。

质美：九龙壁石肌理缜密、细腻，光洁度极高，温润感极强；叩之声如金玉，观之泽似油脂。热力 a new pet in the family of appreciation stones. Their beauty is impregnated in the following aspecgts.

Texture: The stone has a dense and fine texture, with a soft and smooth touch. When knocked, it sounds like gold or jade. The stone has a gleaming luster. The color tone and patterns formed by thermal geology and iron substance bestowed the Jiulong Stones with characteristics similar to that of jade.

- 碧龛晓曙（九龙壁石）
  Dawn in the Green Niche (Jiulong Stone)

- 海峰突窥月（九龙壁石）
  Moon on the Sea (Jiulong Stone)

地质及含铁质成分共同形成的色调和花纹，使其具有碧玉类玉石的特征。

形美：凡摩氏硬度达7以上的高硬度石头，在造型上很难有大的变化，但九龙壁石却是个例外，造型千变万化，令人称绝。

色美：组成九龙壁石的矿物有透辉石、石英、绿帘石、黝帘石及少量的硅灰石、透闪石、石榴子石、长石、方解石等，这些矿物含量的变化造就了九龙壁石的缤纷色彩。红、橙、黄、绿、蓝、靛、紫七色俱全，有单色，也有多色组合，以翠绿、古铜、墨玉、五彩为贵。

纹美：九龙壁石中的纹理图案对比强烈，象形逼真，天然成趣，山峦、江河、花卉、树木、原野、人物、动物、文字皆具美感。

意美：九龙壁石体现了自然山水之美，意境深远，颇具画意。

Shape: Generally speaking, it is very difficult for stones with a hardness grade exceeding Mohs rating of 7 to have beautiful and diverse shapes. Yet the Jiulong Stone is an exception. It is famous for its incredible shapes which model myriads of changes.

Color: Jiulong Stones are mainly composed of many minerals, such as diopside, quartz, epidote and zoisite. They also have ingredients such as silicon limestone, amphibole, garnet, feldspar and calcite. The varied quantities of these minerals enabled the rich colors of the stones. Jiulong Stones of all spectrum colors can be found: red, orange, yellow, green, blue, indigo and violet; some are single-colored, others are mix colored. Among them, green, bronze, black and multi-colored stones are more precious.

Grain: vivid and strong contrasting patterns on Jiulong Stones are natural points of interest; while images of mountains, rivers, flowers, trees, fields, humans, animals and words all have their own sense of beauty.

Charm: Jiulong Stones embody the natural beauty of mountains and rivers, with profound significance and picturesque presentation.

三江石

三江石因产于广西三江侗族自治县而得名,是国内外久负盛名的名石之一。三江石的品种主要有红彩卵石、紫彩卵石、黄蜡石、象形石、景观石、文字石、画面石、金钱石、梨皮石、砂积石、油石、铁石、黑石等数十个品种。其中红彩卵石有全红、血红、花红、紫红、斑纹红等;紫彩卵石则有红紫、花紫、灰紫等;蜡石有红、黄、白、

Sanjiang Stones

Sanjiang Stones are named after their production area in Sanjiang Minority Autonomous County, Guangxi. As one of the prestigious stones in China and the world, Sanjiang Stones entail a number of varieties including: red pebbles, violet pebbles, yellow alabasters, shaping rocks, landscape rocks, character stones, image stones, golden coin stones, pearl peel stones, sand stones, oil stones, iron stones and black stones. Among them,

- 寿比南山(三江石)
  Longevity (Sanjiang Stone)

- 三江石
  Sanjiang Stone

绿、花等，可谓石色缤纷，争奇斗艳，韵味浓郁。三江石石质坚韧，由于三江石的原岩是广西古老的变质岩系，有硅质岩、碧玉岩、含铁石英岩、脉石英、辉绿岩等。硅质岩、碧玉岩的三江石结构致密，所以使石体显得坚韧，非常符合收藏奇石的硬度标准，不容易风化破碎。三江石石质硬密度大，往往一块不起眼的石头抬起来感觉很重，就算体积不是很大的三江石，往往也需要几个人才抬得起来。

red pebbles can be further divided into subcategories of complete-red, blood-red, mixed-red, violet-red and dapple-red. Violet pebble stones include violet, mixed-violet and gray-violet. Alabaster stones have many colors such as red, yellow, white, green and mixed colors. Indeed the stones deserve to be called the competing beauties! Because Sanjiang Stones were transformed from ancient metamorphic rock series including silicon, jade, iron quartzite, gangue quartzite, diabase rocks, silicon and jade, this kind of stones have very high hardness grade and are very heavy, meeting the criteria for wonder stone collection: such stones are not easily weathered or broken. Usually Sanjiang Stones weigh more than they appear to be because of the high density and large gravity. A small piece might need several persons to lift.

- 疏林秋晚（三江石）
  Forest in Late Autumn (Sanjiang Stone)

黄河石

黄河发源于青海，经甘肃、宁夏、内蒙古、陕西、山西、河南，由山东入渤海，途中穿山越峡，流经岩石坠入河中冲磨为砾石，砾石美者即为黄河观赏石，简称"黄河石"。黄河石一般分为山水景观石、形象石、纹理石、色彩石、生物化石等。黄河石的特点是气势大、姿态各异、纹理对比鲜明，千变万化，色彩艳丽、光泽丰润，线条流畅，形、色、纹、图神韵十足，形象逼真。因各河段山石岩质、矿物成分、自然条件各异，所以形成了不同种类的黄河石。

青海江河源石

青海江河源石产于青海境内的黄河上游主河道，一般体量硕大，

Yellow-river Stones

Originating in Qinghai Province, the Yellow River flows through Gansu, Ningxia, Inner Mongolia, Shaanxi, Shanxi, Henan and finally empties into the Bohai Sea in Shandong Province. The River runs through many mountains and valleys where rocks are ground and polished into pebbles and gravels. These beautiful gravels became "Appreciation Stones from the Yellow River" or "Yellow-river Stones". The stones can be divided into categories of landscape stone, image stone, striated stone, color stone and biological fossils, etc. Yellow-river Stones are featured by their majestic grandeur, incredible shapes, strong contrasting grains, flamboyant colors, rich lusters and fluent lines. Charm is impregnated in the shape, color, grain and image of the stones. Different types of Yellow-river Stones were generated at different sections of the Yellow River which host different types of rocks containing different mineral ingredients and subject to different natural conditions.

- 神龟（江河源石）
  The Mysterious Tortoise (River-source Stone)

● 桃（江河源石）
Peach (River-source Stone)

质地坚密，石肌细润，造型奇特，色泽深沉，多见青、黑、灰色。少见纹理，石表圆滑、柔畅，别有韵致。

兰州黄河石

兰州黄河石主要产于黄河上游刘家峡水库至宁夏青铜峡水库的黄河河道，尤以兰州地槽一段最多，故称"兰州黄河石"。兰州黄河石历史悠久，据宋朝杜绾《云林石谱》中的"兰州石"条介绍："兰州黄河水中产石，有绝大者，纹采可喜，间于群石中得真玉，璞外有黄络，又有如物像，黑青者极温

Qinghai River-source Stones

Qinghai River-source Stones are produced in the main channels of the upper stream of the Yellow River within the territory of Qinghai Province. The stones are usually huge in size, with a firm texture, high density, incredible shapes, smooth surfaces and dark color tones. Most of the stones have blue, black and gray colors. Grains are rarely seen. The stone's smooth surface, tender and fluent lines give expression to its unique charm.

Lanzhou Yellow-river Stones

Lanzhou Yellow-river Stones adopted the name because they are mainly produced in the river beds of the upper stream of the Yellow River between the Liujiaxia Reservoir and the Qingtongxia Reservoir. Since most of the stones were found near Lanzhou, so "Lanzhou" was added to the name as the prefix. Lanzhou Yellow-river Stone enjoys a long history. In the chapter on "Lanzhou Stones" of the book *Yunlin Stone Collection*, Du Wan, a famous scholar of the Song Dynasty introduced the Lanzhou Yellow-river Stone stating "Lanzhou section of the Yellow River produces big and strange stones, with beautiful grains and yellow

- 律（兰州黄河石）
  The Rhythm (Lanzhou Yellow-river Stone)

润，可试金。"

兰州黄河石主要岩质为火成岩、沉积岩、变质岩三种，由于在黄河浊水激流下久经磨砺，逐渐形成质地坚硬、造型生动、线条流畅、图纹美丽、色彩斑斓的天然特色，整体古朴典雅、沉稳凝重，恰与西部地域的自然色彩十分吻合，透着一股质朴的美感。

洛阳黄河石

洛阳黄河石主要产于黄河三门峡至孟津河道弯处、河床低凹处。其石体较大，色彩艳丽，有红、橘

lines, as well as images of animals. The black ones are very tender and sleek and can be used to test gold."

Lanzhou Yellow-river Stones are mainly composed of three types of rocks including igneous rocks, sedimentary rocks and metamorphosed rocks. Having been ground and corroded in muddy and rapid currents of the Yellow River, these stones gradually cultivated their natural characteristics such as hard texture, vivid shapes, fluent lines, beautiful patterns and dazzling colors. The stones as a whole give an air of simplicity and elegance, serenity and dignity, blending perfectly with the natural colors of the western region representing rustic beauty.

Luoyang Yellow-river Stones

Luoyang Yellow-river Stones are mainly produced in the bending and recessed river channels between Sanmenxia and Mengjin. These big stones have flamboyant colors, such as red, yellow, white and black. Patterns on these stones are in contrasting colors and bold designs such as semi-circles, standard circles and crescent circles. Some circles have chromatic colors within, like that of solar eclipse, lunar eclipse, annular solar eclipse and annular lunar eclipse. Some

黄、白、黑等色；图案对比鲜明，有半圆、月牙圆、规范圆，有的圆中色套色，恰似日食、月食、日环食、月环食，佳者在日月周围还点缀有朵朵祥云或彩线涡纹，又称"太阳石""日月星辰石"。其石质光滑细腻，有硅质玛瑙、半透明的玉质、砂岩、石灰岩等十多种。

excellent pieces are dotted with colorful clouds or eddy grains around the sun and the moon. So they are also called "solar stones" or "solar-lunar-star stones". They have sleek and exquisite textures and can be further divided into more than a dozen sub-types such as silicon agate, semi-transparent jade, sand-rock and limestone, etc.

- 清明上河图（洛阳黄河石）
Riverside Scene at Qingming Festival (Luoyang Yellow-river Stone)

- 敦煌壁画（洛阳黄河石）
Dunhuang Fresco (Luoyang Yellow-river Stone)

- 巨人对话（洛阳黄河石）
Dialogue between Giants (Luoyang Yellow-river Stone)

泰山石

泰山石,因产于山东省泰山地区而得名。泰山地区是中国古老的太古代变质岩系出露石地区之一,主要是片麻岩、片岩及复杂的变质岩类。神秘而粗犷的泰山石有25亿年历史,是最古老的岩石之一。自秦始皇登临泰岱立石颂德,其后历朝天子,封禅而置石者,不乏其人。

泰山石颜色单一,灰白底子上常有星点般角闪石,体积巨大,常用于园林、楼前置石。通常泰山石

Taishan Stones

Taishan Stones are named after their place of origin in Mount Taishan area, which is one of the areas where Archean metamorphic rock series were exposed. Major rock types include gneiss, schist and complex metamorphic stones. With a history of 2.5 billion years, Taishan Stones are one of the oldest rough and mysterious stones in China. Since the first Emperor of the Qin Dynasty erected a stone tablet to record his merits and virtues in Mount Taishan, many emperors in the ensuing dynasties followed suit.

• 泰山文字石
Taishan Stone Tablet with Inscriptions

的外表以不规则卵形居多，结晶颗粒较粗，纹理清晰，画面突出，色调对比强烈，具阳刚豪放气概。

泰山石不仅在石文化中因内涵丰富而珍贵，同时也具有很高的观赏价值。泰山石首先要求形奇。形奇，即标新立异，或丑极反美，瘦可见骨，或漏中见奥，透隐豁然。元代书画大家赵孟頫配图诗中写道："粤从混沌元气判，自然凝结非琢磨，人间奇物不易得，一见大呼争摩挲。"由此知道泰山石之奇美。上品泰山石的色泽光润亮泽，纹理清晰、细密并富于变化，石质

Taishan Stones are in single color with hornblendes dotting the grayish white stone body. Because of their huge size, they are always placed in gardens or in front of houses. The stones are mostly in irregular egg-shape, with coarse crystal particles, clear grains, outstanding patterns and strong contrasting color tones, showing the spirit of masculinity and bravery.

Taishan Stones are precious not only because of the rich cultural connotation, but also their high appreciation value. The first characteristic of Taishan Stones is its grotesque shape. This is to say that the shapes are so unusually ugly that they become extraordinary. Some stones are extremely ugly and bony; others have hidden caves that contain mystery. In his poem inscribed on a painting, Zhao Mengfu wrote "The mystery results from chaos. It is naturally formed rather than artificially carved. The stones are rarely seen in the world, no wonder people rush over to touch and feel it." From this poem, we can understand the unique value of Taishan Stones. Top-grade Taishan Stones have bright colors with a soft sheen, clear grains, hard and dense texture, evenly distributed crystal particles with naturally smooth surfaces.

• 天马行空（泰山石）
Flying Horse in the Sky (Taishan Stone)

坚韧细密，结晶体分布均匀，并有自然光润感。通常，泰山花岗石卵石质地坚韧，阳光下可见均匀结晶体，往往光润度较低；泰山花玉石、燕子石质地细密，坚韧度低于花岗石，但光润度较高；景观石的坚韧不如花岗石，光润不如玉石，但形体变化奇特。

In general, Taishan Granite Pebbles enjoy a higher hardness level but a lower sleek level. Their symmetrical crystals could be seen under the sunshine. Taishan Color Jade and Swallow Stones have fine texture, lower hardness level but higher sleek level. Taishan Landscape Stones are not as hard as granite, neither are they as sleek as jade, but their shapes are more grotesque and impressive.

● 泰山风光 (图片提供：图虫创意)
Scenery of Mount Taishan

燕山石

燕山石，主要是指产于北京市燕山的奇石。燕山地质构造异常复杂，既有喀斯特地貌出产的钟乳石，也有石英岩、千页岩、海蚀岩和板岩，所以观赏石资源非常丰富。燕山出产的观赏石被统称为"燕山石"，种类有数十种之多。其颜色多见有灰青、褐色、赭红夹青、纯白、青灰夹黄等，纹理逼真，富有变化，质感古朴，光泽凝重，形态极为丰富。金海石和轩辕石是燕山石中最著名的品种。

Yanshan Stones

Yanshan Stones refer to wonder stones found mainly in the Yanshan Mountain area, Beijing City. Due to the complex geological structure of the Yanshan Mountain, both stalactites found in Karst landforms and other rocks such as quartzite, shale, marine-abrasion rocks and slates could be found in the mountain, contributing to the rich resources of appreciation stones. Appreciation stones originated from the Yanshan Mountain area all bear the generic label of "Yanshan Stones". But there are more than a dozen sub-varieties. The most commonly seen colors of Yanshan Stones include gray, brown, pale red mixed with green, white, and gray mixed with yellow, etc. These stones have vivid and varied grains, simple textures, dignified luster and diverse shapes. Gold Sea Stones and Xuan-Yuan Stones are the most famous types of Yanshan Stones.

- "碧水青峰"盆景（燕山石）（图片提供：FOTOE）
Bonsai Landscape "Clear Water and Green Hills" (Yanshan Stone)

金海石

金海石产于北京东部的金海湖一带，因此而得名。金海石的原岩是十几亿年前远古代石英岩，在1.5

- 黄河之水天上来（金海石）
Water from Heaven—the Yellow River (Gold Sea Stone)

- 金海石
Gold Sea Stones

Golden Sea Stones

Golden Sea Stones are named after their place of origin in the Golden Sea Lake area, to the east of Beijing. The protolith of Golden Sea Stones is quartzite in ancient times one billion years ago. The quartzite was infiltrated and soaked by iron and manganese ore fluids in volcano magma some 150 million years ago. Then high ferric ions and low ferric ions were separately distributed. Through long years of weathering, these stones dropped into rivers. Having been brushed and ground by rivers, this quartzite turned into brownish-yellow, crimson, dark brown and violet pebbles. Golden Sea Stones have varied grains, like majestic and splendid landscape paintings. Different from the obscure grains and patterns on other types of stones, grains on Golden Sea Stones are very clear and refined. The stones look like Chinese paintings with color tones between meticulous sketch work and romantic brush work. Layout of the patterns are very appropriate, with profound backgrounds, rich layers, prioritized themes and echoing between major and supporting elements, overlapping mountain peaks, giving full expression to the magnificence of the five major Chinese mountains, such as

亿年前受火山岩浆中所含铁和锰矿液浸染、渗透而使高价铁和低价铁间隔分布，经漫长的风化而碎落江河，再经河水冲刷磨砺成褐黄色、暗红色、黑褐色和红紫色等色彩丰富的景观卵石。金海石纹理变化万千，其绝大多数呈现群峦叠翠、气势磅礴的山水画卷。它区别于一般纹理景观石的朦胧图案，纹理感觉特别清晰细腻，浓淡相宜介于中国画的工笔和写意之间，画面虚实相生疏密得体，背景深远天际朦胧，主次呼应层次丰富，峰峦突兀重重叠叠，展现了五岳之雄伟、天台之险峻、峨嵋之奇幽、山村之幽静，被誉为"画在石上的中国画"。

轩辕石

轩辕石产于北京平谷东北燕山南麓，是一种铁质岩或铁质灰岩，形成于8亿年前的元古代震旦纪。其质为硅质灰岩，孕育于红黏土中；因海水退却后长期沉积于地下，经不断侵蚀，其表面紧附一层红黏土；内质被浸染而呈浅灰微绿色；肌理缜密，含铁量高。因其最初发现于庙山，山上有座轩辕庙，因此取名"轩辕石"。

轩辕石质地纯正、细腻坚实；

the steep Heavenly Terrace Mountain, the tranquil Emei Mountain and the silent rural villages. Golden Sea Stones have been honored as "Chinese paintings on stones".

Xuanyuan Stones

Xuanyuan Stones were found at the southern foot of the Yanshan Mountain, in Pinggu County, Beijing. As a member of the ferruginous rock or ferruginous limestone family, Xuanyuan Stones were formed during the late Sinian period of the Proterozoic Era some 800 million years ago. Because the stone was formed in red clay, it stayed as sediments underground for a long time after the sea water receded. Continued corrosion caused the red clay wrapping the surface of the limestone to penetrate into the stone body, tinting the internal texture. The stone shows a light gray and slightly green color, with refined grains and high iron content. Because the stone was first found in Miaoshan Mountain where there was a Xuanyuan Temple, it was named "Xuanyuan Stone".

Xuanyuan Stones are pure in texture, refined and solid; their shapes are strange and rich in connotation; they produce clear and crisp sounds when knocked;

造型奇特，内涵丰富；击之有声，清脆悦耳；石色稳重大方；肌理自然多样，雄健不失清秀，粗犷不失细腻，气质风度非同一般。轩辕石大者可置于园林庭院，小者可置于文房几案，自然成景。

天峨石

天峨石产自广西红水河上游河段，属纹理石，包括平纹石和凸纹石（也称浮雕石），都是河床中的卵石类。平纹石纹理多为浅褐色、黄褐色、褐棕色，凸纹石的纹理较深，多

their colors are steady and graceful. Due to their diverse and natural grains, they are strong and beautiful, rough and exquisite, with extraordinary charm. The large piece Xuanyuan Stones can be placed in gardens to create sceneries and the small pieces can be placed on study room desks to form a miniature natural landscape for appreciation.

Tian'e Stones

Tian'e Stones originated from the upper stream of the Hongshui River, Guangxi Zhuang Autonomous Region. Falling

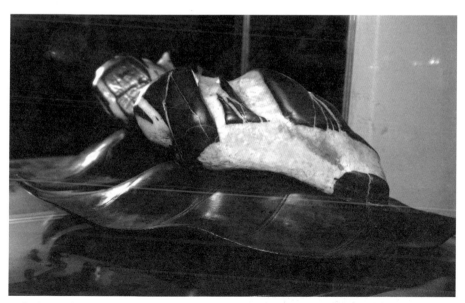

- 春蚕（天峨石）
  Spring Silkworm (Tian'e Stone)

为深褐色、褐黑色、棕黑色等。

天峨石的纹理往往形成一些文字、人物或景观。由于纹理色深，色彩反差大，观之有浮雕画的感觉，而其凸起的纹路则富于粗细、曲直、长短千变万化，细细品赏，内涵丰富。

菊花石

菊花石也称"石菊花"，广泛分布于南方的湖南、广西、江苏、贵州等地，其中以湖南浏阳出产的菊花石最为有名。迄今为止，菊花石的发现已有数百年。据清代文人张尚瑗在《石里杂识》中的记载：

under the striated stone category with flat and convex grains (also called relief stones), Tian'e stones are essentially pebbles in riverbeds. Most of the grains on flat striated stones show light brown, yellowish brown or dark brown colors while grains on convex striated stones show dark brown or brown-black colors.

Tian'e Stones' grains usually form some characters, humans or scenes. Because of the darker color grains and contrasting colors, the stones tend to impress people with a sense of relief carving. The convex grains are so varied with some being thick and others thin, some bent and others straight, some long and others short. Only through careful contemplation, one can perceive the rich connotation of the stone.

Chrysanthemum Stones

Chrysanthemum Stones are also known as "Stone Chrysanthemums", which are widely distributed in southern China regions, such as Hunan, Guangxi,

• 苏州拙政园的菊花石
Chrysanthemum Stones in the Humble Administrator's Garden, Suzhou

"吉水永丰有石，青质而黄章，章为菊花，金英粲然如画。"菊花石上布满惟妙惟肖的菊花状图案。这不是花的化石，而是柱状、针状、纤维状矿物呈放射状或束状排列的集合体，由天然的天青石或异质同象的方解石矿物构成，呈放射状对称分布组成白色花瓣；花瓣中心由近似圆形的黑色燧石构成花蕊，像盛开的菊花。

最常见的菊花石是黑底白花，菊花花瓣为多层状，具立体感。花

● 菊花石
Chrysanthemum Stones

Jiangsu and Guizhou. Among them, Chrysanthemum Stones that originated in Liuyang County, Hunan Province are the most famous ones. Chrysanthemum Stones were found several hundred years ago. In his book *Miscellaneous Knowledge on Stones*, Zhang Shangyuan, a scholar of the Qing Dynasty, wrote: "Jishui and Yongfeng Counties produce beautiful stones with green texture and yellow grains. The grains look like beautiful chrysanthemums." These stones are covered with vivid chrysanthemum patterns. But they are not flower fossils, rather they are collections of natural celestite or calcite minerals in columnar, needle-like and fiber-like shapes. The radially and symmetrically distributed minerals form the white petals and roundish black flint stones near the center form the pistils. Indeed these stones look like blooming chrysanthemums.

Most of the commonly seen Chrysanthemum Stones have a black background with white flower patterns. The petals are in many layers, giving a three-dimensional effect. The flowers vary in size and shape including silk balls, phoenix tails or butterflies. White and bright chrysanthemums against the black background, occasionally dotted

朵大小不一，花形各异，有绣球状、凤尾状、蝴蝶状等。白色晶莹的菊花，陪衬着黑色基质岩石的底色，黑白分明，灰岩中偶尔含有蜒类、腕足类珊瑚化石，给菊花石增添了生命活力。

天然菊花石因产量稀少，比珠玉宝石类更珍贵，故被誉为"全球一绝"。对于菊花石的鉴赏，主要从以下三个方面入手。

by fossils of dragonfly-type creatures or brachiopod-type coral fossils, all these features render the Chrysanthemum Stones much vigor and vitality.

Because of the small production quantity, natural Chrysanthemum Stones are more precious than jade and pearls. Therefore, they are also reputed as the "unique stone in the world". When appreciating Chrysanthemum Stones, we should pay attention to the following three aspects:

Shape: The more the shape is closer to a chrysanthemum, the more beautiful the stone is. Chrysanthemum patterns on the stones take diverse forms and

● **菊花怒放（黄石菊花石）**
除了湖南省浏阳地区以外，中国南方许多省区也出产菊花石。产于湖北省黄石地区的菊花石特点鲜明，菊花图案有白、黄、红、黑等颜色，不是单一的白色，色彩可人，而且花蕊和花瓣有凸有凹，边界清晰分明，富有立体感。

Chrysanthemum in Full Blossom (Huangshi Chrysanthemum Stone)
Besides the Liuyang area of Hunan Province, Chrysanthemum Stones were also found in many other provinces and regions in southern China. For instance, Chrysanthemum Stones found in Huangshi area of Hubei Province distinguished themselves from those found in other areas: instead of single white color petals, chrysanthemum patterns on Huangshi stones are of many colors including white, yellow, red and black. In addition, the pistils and petals are presented on different planes with clear borders, some bulging, others sunken, showing strong three-dimensional effects.

形，即菊花石的外形，其形状越是接近菊花越美。菊花石花形千姿百态，由花蕊和花瓣两部分组成，其形态特点是以花蕊为中心，向四周作三维放射状延伸，如同自然界的菊花一样千变万化。

质，菊花石的石质越细腻越好。通常，菊花石的花瓣要大且细腻，但并不像真花那样线条流畅优美，而是呈独特的放射状。

色，指菊花石的色泽。通常其花朵颜色与基质色差较大为美，这样给人一种立体感，从而增强了观赏性。

postures. Basically, they are composed of pistils and petals. The petals radiate from the central pistils in three-dimensional arrays, resembling real chrysanthemums in the natural world.

Texture: The finer the stone texture, the better the chrysanthemum stone. Generally speaking, the petals should be big and exquisite. However, unlike real flower petals that are free-flowing, petals of Stone Chrysanthemums show unique radiating patterns.

Color: This refers to the color of the chrysanthemum pattern on the stone. In general, stones with strong contrasting colors between the background and the flower patterns are considered top-grade pieces, as they give the viewer a sense of three-dimensional beauty for appreciation.

牡丹石

牡丹石的底色为黑色，石体上分布了很多或白或绿的晶状体，宛如一朵朵国色天香的牡丹，因而得名。牡丹石产于河南省寇店镇，深藏在万安山的褶皱中。有些人误以为牡丹石是"花"的化石。其实不然，两者之间的区别主要表现在牡丹石的"花"是由矿物成分组成，但难嵌合得严密，侧面呈现较强的立体感；而植物的花、叶、茎是很

Peony Stones

Peony Stones got their name because of the many white and green crystals on the black stone body, like blossoms of beautiful peonies. Peony Stones are produced at Koudian Town of Henan Province. The stones are deeply concealed in the cracks of Wan'an Mountain. Some people erroneously believe that Peony Stones are fossils of "flowers". This is not true. "Flowers" on Peony Stones are formed by minerals that are difficult to be embedded completely into the stone body. So the side profile of the flowers shows strong third-dimension

- 河南洛阳龙门石窟的佛手牡丹石 (图片提供: FOTOE)
Buddha's palm Peony Stone in the Dragon Gate Grotto, Luoyang, Henan Province

- 牡丹石
Peony Stones

柔软的，成为化石后只是在岩层表面压成印痕，很难呈现立体感，只要稍加比较，就可看出明显的区别。

牡丹石的珍贵之处不仅在于其美，更在于稀少。专家经考察后得出的结论是，牡丹石在世界上独一无二，只在中国洛阳有极少的储量，且是不可再生的资源。

枣花石

枣花石主要产于山东省临朐县石家河乡宝畔台村和白沙乡石瓮沟村。枣花石是一种泥质岩，底色为灰白色、黄褐色，石上布满白黄色星点纹理，恰似盛开的成串枣花。

● 枣花石 (图片提供：FOTOE)
Jujube Flower Stone

effects. In contrast, flowers, leaves and stems are soft and have been pressed into prints only in the process of petrification. It is difficult to see three-dimensional effects on flower fossils. By comparing the two, it is not difficult to see the differences between the two "flowers".

Peony Stones are precious not only because of their unique beauty, but also because of their scarcity. According to study findings by specialists, Peony Stones are unique in China with very few reserves in Luoyang. In addition, these stones are non-renewable natural resources.

Jujube Flower Stones

Jujube Flower Stones are mainly produced at Shiwenggou Village of Baisha Town and Baopantai Village of Shijiahe Town, Linqu County, Shandong Province. Jujube Flower Stone is a member of the argillaceous rock family. The stone's background is usually in light gray or tanned color, with light yellow spots dotting the stone body, like bunches of blooming jujube flowers. The stone's texture is hard and fine, with clear and beautiful grains. After the raw stone is excavated, incised and polished, it reveals beautiful natural grains and

枣花石质地坚硬细润，纹理清晰美观，挖掘出土后，经切割打磨，即露出枣花、树木、水草、湖泊等图案，也有古松、梅竹、山泉、溪流等花纹，呈象优美、古朴自然。此石可加工成各种大小石屏，镶上底座，陈设于几案或装点厅堂，潇洒雅观，颇具韵味。

丹麻石

丹麻石，因产自昆仑山东麓的青海省湟中县丹麻乡而得名。丹麻石属沉积岩类，主要矿物成分为方

patterns like jujube flowers, trees, weeds, lakes, ancient pines, bamboos and plum blossoms, streams and brooks, etc. Jujube Flower Stones can be processed into various sized screens, which can then be placed on tables or in halls together with a stand. They are elegant, graceful and charming items for decoration and appreciation.

Danma Stones

Danma Stones are named after their place of origin in Danma Town, Huangzhong County, Qinghai Province, which is located in the east of the Kunlun Mountain. Danma Stones belong to the sedimentary rock family, whose main ingredients are calcite, limonite, dolomite and clay. The texture is exquisite and bright. Because of the rich ferric element, the stones show yellow color while the grains show yellow, light yellow, golden yellow, brown, black and white colors, etc. Due to regular sediments of various combinations and long years of pressing and modification, various beautiful grains were formed on Danma Stones. Some of the commonly seen grains include veins,

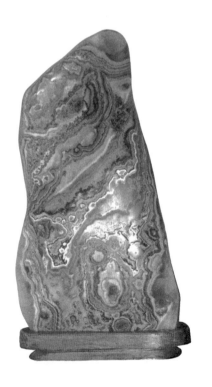

- 丹麻石
Danma Stone

解石、褐铁矿、白云石及黏土等。丹麻石质地细腻莹洁，因其富含铁元素而多呈黄色，其纹理色泽有金黄、浅黄、土黄、棕、褐、黑、白等。由于各种组合有规律的沉积，又经漫长地质年代的挤压、改造，丹麻石形成了各种各样的美丽纹理，常见的有脉状、带状、波纹状、点簇状、斑簇状等等，花团锦簇，美不胜收。而且丹麻石的天然纹理布满整个石体。剖开任何一块丹麻石，在任何一个层面上，都可见到美丽的纹理，一般是以连续纹样出现，如团团簇簇的花、层层叠叠的云、连续不断的浪等。

天景石

天景石，因清代原产于山东省曲阜的尼山一带，又称"尼山石"，20世纪80年代末又于山东费县城南的天井汪村被再度发现，取"天井"谐音名为"天景石"。天景石属沉积岩类之泥灰岩，略有变质，微粒结构，质地细腻坚硬，色彩艳丽，纹理清晰，石上纹彩主要是锰的氧化物侵入沉积而成的黑色以及铁的氧化物沉积形成的赭红、

belts, ripples, point clusters and dot clusters, etc. In addition, natural grains penetrate the whole body of Danma stone. If you cut any Danma stone, you could see beautiful grains on any layer of the stone. The grains usually appear as continuous patterns, like clusters of flowers, overlapping clouds and surging waves.

Tianjing Stones

Tianjing Stones were originally found in Nishan Mountain, Qufu City, Shandong Province during the Qing Dynasty. So it was also called "Nishan Stones". In the late 1980s, the stone was re-discovered in Tianjingwang Village, Feixian County, Shandong Province. So the stones were renamed Tianjing Stones. These stones belong to argillaceous rock of the sedimentary rock family with slight variation. They are of particulate structure, hard and exquisite texture, flamboyant colors and clear grains. Grains on the stones are mainly composed of black color which is the intruded and deposited manganic oxides and dark red and tanned colors which are the deposited ferrous oxides. The stone's background colors are generally soft

黄褐色，底色一般为柔和的绢黄色，还有天蓝、绛红、青灰等色彩。经过打磨上光的天景石，几乎件件都是近乎完美的彩色山水画。有的画面上不仅有黄褐色的山岭峰峦及原野，黑色的丛林和花草，还有蓝色的天空，有的在蓝天之上还飘有红云，一块石头一种不同的景象，其色彩之协调，景物之逼真，层次之分明，令人拍案叫绝，回味无穷。

yellow, sky blue, dark red and gray, etc. Almost every piece of polished Tianjing Stone is a near perfect colorful landscape painting with mountains and rivers. On some of the stones, there are not only brown mountains and peaks, green fields, but also black forests and grass or blue sky. Some have red clouds in the sky. Each stone has its own distinct pattern with harmonious colors, vivid images and clear hierarchy structure, triggering great fascination and indelible memory.

● 旭日东升（天景石）（图片提供：FOTOE）
Sun Rise in the East (Tianjing Stone)

蜡石

蜡石，因石体表层有蜡状的质感、色感、光感而得名；另有说法认为，古代称柬埔寨为"真蜡国"，该国向明朝皇帝上贡过一块极品黄蜡石，"蜡石"之名因此而来。

蜡石属硅化安山岩或砂岩，主要成分为石英，油状蜡质的表层为低温熔物，坚韧性强，硬度与翡翠相似，是中国传统奇石中质地

Alabasters

Alabasters are named after the wax-like texture, color and luster on their surface. Another version of the stone is that a kingdom called the "True Wax Kingdom", offered an excellent piece of alabaster as a tribute to the emperor of the Ming Dynasty. This kingdom is now Cambodia. Hence the name "Alabaster" was adopted as name of the stone.

Alabaster is a silicon andesite or sandstone, whose main ingredient is quartz. Its wax surface is a layer of low-temperature fusion, which is as hard as jadeite. Alabaster is one of the hardest stones in the traditional Chinese wonder stone family. As most alabasters are

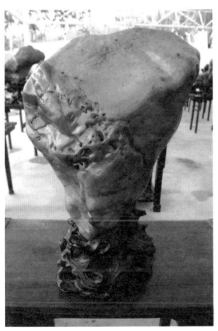

- 圣火（黄蜡石）
  The Holy Flame (Yellow Alabaster)

- 九州雄风（黄蜡石）
  The Majestic Style (Yellow Alabaster)

最为坚硬致密的石种之一。蜡石以黄色最为常见，因此又称为"黄蜡石"，因其在形成过程中渗入的矿物不同，又有红蜡石、绿蜡石、白蜡石、黑蜡石、彩蜡石等品种。

由于其中二氧化硅的纯度和石英体颗粒的大小、表层熔融的情况不同，蜡石又分冻蜡、胶蜡、晶蜡、细蜡和粗蜡等。冻蜡是蜡石中最优秀的品种，俗称"年糕冻"；

yellow, they are also called the "Yellow Alabaster". Due to the different minerals infiltrated during the formation process, other colored alabasters such as red, green, white, black and multi-color also exist.

Alabasters can be divided into subcategories including Jelly, Glue, Crystal, Fine and Rough Alabasters according to differences between silicon dioxide's purity, the size of the quartz particles and the state of surface fusion. Among them, Jelly Alabaster is the best variety. It is also called "Jelly Cake". Glue Alabasters take the form of glue and are slightly transparent. The smooth surface is as sleek as that of Jelly Alabasters. Glue Alabasters have a very fine texture with oil droplet-like or wax-like condensate on the surface. Crystal Alabasters contain crystal-like substances in the surface crannies or holes. The more crystal content the stone contains, the more valuable the stone is. Although Fine Alabasters are opaque, their texture is fine and sleek and the surface is smooth. They have good value for appreciation and fondling. Rough Alabasters are opaque with coarse texture and dull color. Only those with grotesque shapes could have values for appreciation.

● 黄蜡石
Yellow Alabaster

胶蜡呈胶状体，微透光，油润光洁可与冻蜡媲美，石质颇佳，表面常呈油滴状、蜡凝状；晶蜡是蜡石的表面缝隙或空洞之处夹杂有水晶状物质，所附水晶体越多，观赏价值越高；细蜡不透光，但质地细腻油亮，表层平滑，手感好，有赏玩价值；而粗蜡不透光，质粗色暗，手感差，只有造型特别者才具观赏价值。

蜡石主要产于广东和广西，尤

Alabasters are mainly produced in Guangdong and Guangxi. Those from Chaozhou in Guangdong and those from Hezhou in Guangxi boost the best quality.

The history of Alabasters as appreciation stones dates back to the Ming Dynasty. Those alabaster pieces with beautiful colors such as yellow, sleek surfaces could be appreciated and fondled with. Top-grade alabasters are

● 五彩蜡石
Multi-color Alabaster

● 黄蜡石
Yellow Alabaster

以广东潮州和广西贺州的蜡石质地最好。

　　蜡石作为观赏石已有较长历史，最早可追溯到明代。其色泽美丽，油腻润泽，金黄红艳，可观赏，可把玩，顶级蜡石更是难得。黄蜡石之所以能成为名贵观赏石，除其具备湿、润、密、透、凝、腻外，其主"色相"以黄色为基本色是重要因素。黄蜡石的黄色，为亮丽的暖色调，黄色自古以来就是权力与富贵的象征，历代皇帝所穿龙袍是以黄色为主色调，所以人们喜欢黄蜡石有着历史的渊源。

very scarce in the world. An important reason for Yellow Alabasters to become precious appreciation stones could be attributed to their "primary yellow color", in addition to the other characteristics such as their moisture, luster, density, transparency, congealment and smoothness. Yellow is a bright and warm color and it has been used as the symbol of power and wealth in ancient China. Many emperors in ancient China picked yellow color as the primary color of their imperial dragon gowns. So people's love for Yellow Alabasters is a natural outcome of historic precipitation.

石中新贵黄龙玉

　　黄龙玉，又称"龙黄石"，产自云南省保山市龙陵县小黑山自然保护区的龙江边，属于一种石英质的玉石，人称"云南黄蜡石"。其主色调为黄色，兼有羊脂白、青白、红、黑、灰、绿等色，有"黄如金、红如血、绿如翠、白如冰、乌如墨"之称。云南黄蜡石石质细润，色泽金黄，块型硕大，变化丰富，有很高的观赏价值。

Yellow Dragon Jade — A New Star to the Wonder Stone Family

Yellow Dragon Jade is produced on the banks of the Dragon River, Xiaoheishan Natural Reserve, Longling County, Baoshan City, Yunnan Province. Also known as the "Yellow Alabaster in Yunnan", Yellow Dragon Jade is a type of quartz jade. Its main color is yellow. Other colors include suet white, light gray, red, black, gray and green, etc. As vividly described by some people that the colors of these stones are "as yellow as gold, as red as blood, as green as jadeite, as white as ice and as black as ink". Yellow alabasters found in Yunnan have exquisite texture, golden yellow color, huge size and rich variations. They bore high values as appreciation stones.

- 福寿满堂（黄龙玉）
  A House of Happiness and Longevity (Yellow Dragon Jade)

大理石

大理石在古代称为"云石"，原指产于云南大理的一种白色带有黑花纹的石灰岩。在地壳的内力作用下，原来的各类岩石发生质的变化，形成了新的岩石，大理石主要由方解石、蛇纹石和白云石组成。

大理石自古以来受到人们的喜爱，不仅由于它质地纯洁、结构细匀，更在于石上变化无穷的美丽图案。而在这些千变万化的画面中，尤以酷似山水画者最多，无论是峰

Marbles

Marbles were called "Cloud Stones" in ancient times in China, referring to mainly a type of limestone produced in Dali, Yunnan Province. The stone has white texture and black grains. Composed mainly of calcite, toxoite and dolomite, marbles were formed through internal dynamics of the earth crust where all kinds of rocks were substantially transformed into new rocks.

Marbles have always been appreciated by the Chinese people, not only because of

• 红木灵芝纹嵌大理石靠背太师椅（清）
Red-wood Marble Backrest Chair (Qing Dynasty,1616-1911)

• 红木嵌大理石座屏（清）
Marble Table Screen on Red-wood Setting (Qing Dynasty, 1616-1911)

峦岩壑、瀑布溪涧、峻坂峭壁、长江大河，无一不是大自然精妙绝伦的杰作。

明清时期，大理石受到人们的喜爱，当时的紫檀、黄花梨等各类家具，几乎无一不用大理石装饰，如几案的台面，椅榻的靠背等。人们还用大小不等的大理石片制成各种插屏、挂屏，深受皇室贵戚、文人雅士的青睐。明代著名的旅行家徐霞客曾称赞大理石屏之美："造物之愈出愈奇，从此丹青一家皆为俗笔，而画苑可废矣。"清代著名书画家郑板桥也曾称赞大理石屏的画面之美。

- 苏州网师园内大理石挂屏
Wall Marble Screen in Master-of-nets Garden, Suzhou

their pure texture and exquisite structure, but also because of their varied and beautiful patterns. Most patterns mimic landscape paintings of mountains and rivers, peaks and valleys, waterfalls and streams, gorges and crags. All the pieces are masterworks of nature.

Marbles were very popular during the Ming and Qing dynasties, when much high profile furniture made of red sandalwood or yellow pear wood used marbles as decorations to complement the furniture. For instance, marbles were made into desktops and chair backrests. People also used different-sized marbles to make desk screens and wall screens, both were highly adored by members of the imperial families, senior officials and scholars. Xu Xiake, a famous traveler of the Ming Dynasty, once praised the marble screens with the following wording: "Nature has created more and more wonders! Compared to this, paintings done by artists are so vulgar, and painting schools should be closed." Another famous painter of the Qing Dynasty Zheng Banqiao once praised the beauty of marble, either.

玛瑙石

玛瑙石是指具有纹带构造的玉髓，是一种胶状矿物，主要成分为二氧化硅。玛瑙主要产于火山岩裂隙及空洞中，也有的产于沉积岩层中，与水晶、碧玉等一样都是一种

- 收获（葡萄玛瑙石）

葡萄玛瑙石产于内蒙古阿拉善盟苏宏图一带，石质坚硬晶莹，色彩绚丽，呈浅红至深紫等色，半透明，石上通体满布色彩斑斓、大小不一的珠状玛瑙小球，互相堆积，犹如串串葡萄。

Harvest (Grape Agate)

Grape Agates are produced in Suhongtu area, Alxa League, Inner Mongolia. The stone's texture is hard and crystal, with bright colors ranging from light red to deep purple. The semi-transparent stone body is covered by colorful agate balls of various sizes, like clusters of grapes.

Agates

As a colloidal mineral, agate refers to chalcedonies with vein-belt structure. The main ingredient of agate is silicon dioxide. Agate Stones are mainly formed in volcanic lava crannies and caves or sedimentary layers. Same as crystals and jades, agates are quartz minerals in substance, but they are harder than crystals. Agates could be formed in all strata of geologic history, in both igneous rocks and sedimentary rocks. Agates are widely distributed and produced in almost all provinces and regions of China,

- 水胆玛瑙

水胆玛瑙，就是自然形成的玛瑙中包裹有天然形成的水，摇晃时汩汩有声，以"胆"大"水"多为佳。

Agate Enhydritic

Agate enhydritics refer to natural Agate Stones with water trapped in pockets inside its body. When shaken, the water produces a gurgling sound. Agate enhydritic stones with larger inside "pockets" and more trapped "water" are considered the superior quality.

● 霞光地带（玛瑙石）
Morning Glow (Agate Stone)

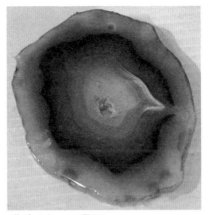
● 荷塘月色（玛瑙石）
Moon-light over the Pond (Agate Stone)

石英矿，其硬度超过水晶。在地质历史的各个地层中，无论是火成岩还是沉积岩都能形成玛瑙。中国的玛瑙产地分布很广泛，几乎各省都有，黑龙江、辽宁、湖北等地最多。

玛瑙是人类最早利用的宝石材料之一，在东方，它是七宝、七珍之一。在中国，古人把红色的玛瑙称为"赤玉"，或称为"琼"。玛瑙由于纹带美丽，自古就被人们用作装饰，在出土的玉器中，常见成串的玛瑙球。

玛瑙本身具有坚硬、质感细腻、形状各异、光洁度高、色彩丰富等特点，除了作为雕琢工艺品的上等材料之外，也常常作为观赏石被人们收藏。

with Heilongjiang, Liaoning and Hubei provinces boosting the top producers.

Agate is one of the oldest precious stone materials utilized by mankind. In the oriental civilization, agate is regarded as one of the seven treasures. In ancient times, people in China used to call red agate "Red Jade" or "Fine Jade". By virtue of their beautiful veins and patterns, Agate Stones have been used for decoration since ancient times. Among those unearthed jade objects, we often find stringed agate beads.

Agate is featured by hard texture, beautiful grains and patterns, varied shapes, sleek surface and rich colors. In addition to being a high-quality carving material, Agate Stones are also collected by the people as appreciation items.

> 矿物晶体

地球亿万年的地质运动，形成了姿态万千、美丽无比的矿物晶体。这些具有多面体外形的晶体，有着奇特、稀有而且美丽的特性。自古以来，中国人就有将矿物晶体或晶簇供于几案的传统。

水晶

水晶是结晶完好的石英晶体，主要化学成分是二氧化硅，其晶体状态是由六方双锥和六方柱构成的带锥头的六方体。水晶晶簇的颜色多种多样，有无色透明或乳白色半透明的白水晶，有紫色的紫晶，烟灰色的烟晶，茶褐色的茶晶，黄色的黄水晶，玫瑰色的蔷薇水晶（芙蓉石）等。

水晶大多呈单晶晶簇产出，

> Mineral Crystals

Billions of years of geological movements have generated diverse and beautiful mineral crystals that are in polyhedron shapes, with unusual, rare and beautiful characteristics. Since ancient times, people in China have cultivated the tradition of placing mineral crystals or crystalloid druses on desks and tables as appreciation items.

Crystals

Crystals are complete quartz crystals whose main chemical ingredient is silicon dioxide. Its crystalloid is a hexagon with a pyramid head composed of hexagonal dual-pyramids and hexagonal prisms. Crystal druses show various colors, such as achromatically transparent or white semi-transparent crystals, violet amethysts, smoky gray cairngorm, tea

有时还可能和其他矿物晶簇（如萤石、重晶石、辰砂或镜铁矿等）共生。与其共生的镜铁矿往往呈花瓣状集合体，构成"铁玫瑰"，形态十分美观。水晶晶簇本身也常组成形如菊花的放射状集合体，很受人们的喜爱。中国境内各地几乎都产水晶，而以江苏、贵州、四川、内蒙古、海南等省区较多。

brown smoky crystals, yellow citrines and pink rose crystals (rose quartz), etc.

Most crystals are in monocrystal druses. They sometimes coexist with other mineral druses, such as fluorites, barites, cinnabars or spiegeleisen ores. Such coexisting spiegeleisen ores are in petal aggregations, forming "Iron Roses" with beautiful shapes. Crystal druses themselves often compose chrysanthemum-like radiating aggregations, that are absolutely adored by people. Crystals are produced in many areas in China, especially in Jiangsu, Guizhou, Sichuan, Hainan, Inner Mongolia and other provinces and regions.

- 晶屋藏娇（紫水晶）
  Hiding a Beauty in a Crystal House (Amethyst)

- 水晶与镜铁矿共生矿
  Intergrowth of Crystal and Spiegeleisen Ore

矿物晶体的鉴赏

对于矿物晶体的鉴赏,主要从完整性、奇特性、精美程度三个方面入手。

1.完整性

矿物晶体不但要求整体的完整性,还要求其局部的完整性。完整有两层含义,百分之百的完整和一个大的观赏面的完整,对完整性的要求取决于矿物的自身条件。以孔雀石为例,它是在铜矿的氧化带中风化而成的,块度较大,不可能有百分之百的完整,因此只能采出一个观赏面完整的孔雀石。而对作为观赏石的矿物晶体及晶簇,强调保持天然的状态,一般须有原生岩石作为基底。

2.奇特性

矿物由于在结晶时受到外部条件的影响,它的奇特之处不是传统赏石可比拟的,如造型的奇特、结晶的奇特、共生的奇特、色彩的奇特等。比如几种生成环境完全不同的矿物,可以生长在一起,以水晶和方解石为例,在自然界中,方解石长在水晶上面是很常见的。一块观赏矿物的奇特之处越多,它的价值也就越高。除此之外,观赏矿物的奇特性也指矿物品种的稀有性。

• 石榴籽石与烟晶共生矿
Intergrowth of Garnet and Cairngorm Ore

3.精美程度

一块好的观赏矿物，应该具有良好的视觉效果，颜色艳丽，晶体粗大，透明度高，光泽度强，造型好。由于矿物晶体是天然形成，不是人造的，因此是不可再生的宝贵资源。它们不仅具有地域性、稀有性、奇特性、艺术性、科学性和商品性等特点，还具有观赏、把玩、陈列、收藏和科学研究价值。

Appreciation of Mineral Crystals

Mineral crystals should be appraised from the following three aspects: integrity, special characteristics and refinement.

1. Integrity

Fine mineral crystals have strict requirements on not only the integrity of the whole mineral piece, but also the integrity of the local parts. This has a two-fold significance: one is the completeness of the entire mineral ore; the other is the existence of a full-viewing surface. Mineral crystals' integrity depends on their intrinsic conditions. Taking malachites as an example, since they have been weathered in the oxidation zone of copper mines and are large in size, it is very difficult to get the entire ore piece. Therefore only a piece of malachite with a perfect viewing surface can be mined. Natural mineral crystals and crystal clusters used for decoration and appreciation purposes should keep their natural forms and conditions and they usually come with the original rock as the pedestal.

2. Special Characteristics

In the process of crystallization, minerals were affected by many externalities. So their unusual features overshadow traditional appreciation stones, for instance, their exotic shapes, unique crystals, interesting intergrowth patterns and dazzling colors. Several minerals with different generating environments can co-exist. Taking crystals and calcite as an example, in the natural world, calcite often grows on crystals. The more unique features an appreciation mineral enjoys, the more valuable it becomes. Besides, the unique features of an appreciation mineral also include the rarity and scarcity of the mineral type.

3. Refinement

A fine piece of appreciation mineral should possess fine visual effects, flamboyant colors, big and thick crystals, high transparency, bright luster and elegant shape. As mineral crystals are naturally rather than artificially formed, they are precious non-renewable resources. They are not only featured by characteristics such as regionality, scarcity, uniqueness, artistry, scientific and commercial significance, but also values for appreciation, fondling, display, collection and scientific research.

方解石

方解石系六方晶体，是一种分布广泛的常见矿物晶体，主要化学成分是碳酸钙，有完全的菱面体解理，玻璃光泽，透明至半透明，普通为白色或无色，有时因含其他元素而呈浅黄、浅红、紫、褐黑色等。纯净透明的方解石称为"冰洲石"，具有强烈双折射和完全解理。

方解石晶簇主要产于以碳酸盐岩为围岩的热液脉型矿床中，贵州、广西、云南等省区有大量碳酸盐岩地层分布，岩浆活动不发育，因此其生成温度一般比水晶稍低。方解石晶体形

Calcites

Calcite is a widely distributed and commonly accessible hexagonal mineral crystalloid. Its main chemical ingredient is calcium carbonate. Calcites possess complete rhombus cleavages, glass-like luster, transparent to semi-transparent texture and white or achromatic color. Because of other elements contained, they sometimes show other colors such as light yellow, light red, violet and dark brown, etc. The transparent calcites are called "Iceland Spar", with intense birefringence and complete cleavages.

Calcite druses are mainly generated in hydrothermal nervation deposits around the carbonate rock strata, which are widely distributed in Guizhou, Guangxi, Yunnan and other provinces and

• 方解石、钟乳石
Calcite and Stalactite

• 方解石
Calcite

- 扇面彩晶（方解石）
  A Fan-shaped Colorful Crystal (Calcite)

- 方解石
  Calcite

态多种多样，造型美观，常见的有柱状体、菱面体、板状体、三角面体等。单晶大小可以从几毫米至数十厘米不等。有时方解石晶簇可和金属硫化物晶体（如黄铁矿、闪锌矿）共生，形态更为美丽。

萤石

萤石又称"氟石"，是一种钙的氟化物，单晶呈立方体、八面体、菱形十二面体及聚形。立方体晶面上常有与棱平行的网格状条纹，集合体为粒状、晶簇状、条带状、块状等。单晶大小可由数毫米

regions. Because magmatic actions are not developed, the generation temperature for calcites is lower than that of crystals. Calcites are formed into various beautiful shapes, such as prisms, rhombi, plates and triangles. A single crystal's size varies from several millimeters to several decimeters. When calcite druses co-exist with metal sulfide crystals (for example, iron pyrites and zinc blends), the shapes are more beautiful.

Fluorites

Fluorite, also known as "Fluorspar", is a member of the calcium fluoride family, with single crystals taking cubic,

至几十厘米，矿物晶体大多为半透明至透明，在紫外线照射下出现极强的荧光。

萤石主要产于热液矿脉中。无色透明的萤石晶体产于花岗伟晶岩或萤石脉的晶洞中。纯净的萤石是无色的，但有些由于含微量元素的原因，几乎各种颜色都有，常见的有绿色、紫色、黄色、红色、褐色、灰色等；在同一晶体上也会有多种颜色，通常见到的是紫色和绿色交杂在一起的情况。中国是世界上萤石矿产最多的国家之一，主要产于浙江、湖南、福建等地。

octahedron, rhombus dodecahedron and aggregation forms. Many grid-like stripes are distributed on the surface of the crystal, paraller to the grains. The aggregations are in particles, druses, strips and lumps. A single crystal's size varies from several millimeters to several decimeters. While these mineral crystals usually show semi-transparent to transparent textures, strong fluorescence appears under ultraviolet radiation.

Fluorites are mainly generated in hydrothermal nervation deposits. Achromatically transparent fluorite crystals are generated in druses of granite pegmatite or fluorite wythern. Pure fluorites are achromatic. Sometimes when fluorites contain microelements, they can also show many colors such as green, violet, yellow, red, brown and gray. The same piece of crystalloid could also have mixed colors, and the most common type is purple mixed with green. China is one of the major producers of fluorites. In China most of the fluorite ores are produced in Zhejiang, Hunan and Fujian.

• 萤石
Fluorite

- 萤石
  Fluorite

- 萤石
  Fluorite

孔雀石

孔雀石体属单斜晶系的碳酸盐矿物，因颜色类似蓝孔雀羽毛上美丽的斑点而得名，中国古代称为"绿青""石绿""青琅玕"。其晶体为柱状、针状或纤维状，形状通常呈钟乳状、肾状等，具有玻璃

Malachites

As a member of the monoclinic carbonate mineral family, malachite got its name after its color which is similar to that of the beautiful dots on the feathers of blue peacocks. In ancient China, malachite was also called "Green Blue", "Green Stone" or "Blue Jade and Pearl Tree". While malachite crystals are in cylinder, needle or fiber shapes, malachite stones are of stalactites or kidney shapes. They have glass-like lusters and semi-transparent textures. Generated in the oxide zone of copper deposits,

- 孔雀石
  Malachite

• 孔雀开屏（孔雀石）
A Peacock Showing its Pride (Malachite)

• 大鹏展翅（孔雀石）
A Hawk Spreading its Wings (Malachite)

光泽，半透明。孔雀石产于铜矿床氧化带中，是含铜硫化物氧化的次生产物，常与蓝铜矿、赤铜矿、褐铁矿等共生，可用作寻找原生铜矿的标志。中国海南岛等地盛产孔雀石。孔雀石光彩夺目，姿态万千，是人们收藏观赏的佳品。

刚玉

刚玉的化学组成为氧化铝，其硬度仅次于金刚石。较佳的品种是蓝宝石和红宝石，而与氧化铁及尖晶石的混合物称为"刚玉砂"。天然刚玉一般都含有微量元素杂质，主要有铬、钛、锰、钒等，因

malachite is the secondary product during the oxidation process of copper sulfide. As malachite often co-exists with azurite, cuprite and limonite, it can be used as a sign for potential copper ores. China's Hainan Island is rich in malachite resources. The bright green color and diverse shapes make malachite an excellent choice for appreciation and collection.

Corundum

The chemical composition of corundum is aluminum oxide, whose hardness grade is only next to diamond. The better corundum types are sapphire and ruby. When corundum is mixed with ferric

此使刚玉带有不同颜色，如黄灰、蓝灰、红、蓝、紫、绿、棕、黑色等。刚玉的晶体形态常呈桶状、柱状或板状，晶形大多较完整，具玻璃光泽或金刚光泽。刚玉分布广泛，呈圆柱形晶体、大块体或圆颗粒，产于火成岩、沉积岩及变质岩中。

oxide and spinel, it became "emery". As natural corundum always contains microelements such as chrome, titanium, manganese and yanadium, it shows various colors, such as yellow, gray, blue gray, red, blue, violet, green, brown and black. Corundum crystals are usually in shapes like a bucket, a pole or a panel and the shape of crystals is relatively complete with glass-like or diamond-like lusters. Corundum is widely distributed in the form of cylindrical crystals, bulk or round particles and was formed in igneous, sedimentary and metamorphosed rocks.

- 金嵌红宝石蜻蜓簪（清）

红宝石，是含铬的具有鲜艳红色的透明到半透明的刚玉，是一种非常珍贵的宝石。中国西藏和云南等地已发现红宝石矿床。

Gold Hairpin with Inlaid Ruby Dragonfly (Qing Dynasty, 1616-1911)

Ruby is a red transparent or semi-transparent corundum containing chromium. As an extremely precious stone, ruby deposits have been found in Xizang, Yunnan and other provinces and regions.

- 粉红刚玉
Pink Corundum

电气石

电气石又称"碧玺",是一种硅酸岩矿物,三方晶系,晶体呈柱状。晶体两端晶面不同,柱面常出现纵纹,横断面呈球面三角形;集合体呈棒状、放射状、束针状,也呈致密块状或隐晶块体。

一百多年来,宝石级的电气石晶体一直是经久不衰的高档陈列、收藏品。决定其观赏价值的最重要因素是颜色。红色最好,蓝、绿、黄绿、浅蓝次之,黑色的电气石较差。

Tourmalines

Tourmaline, also known as "Green Seal", belongs to the trigonal branch of the silicate mineral family. The crystalloid is in a cylinder shape. The two ends of the crystalloid have different surfaces. While vertical veins are often seen on the cylinder, the cross-section shows a spherical triangle. The aggregations are in shapes such as stick, radial, needle cluster, as well as in dense mass or cryptocrystalline blocks.

During the past century, tourmaline crystals of gem quality have always been treated as high profile ornaments and collector's favorites. The most decisive factor determining the value of the stone

• 粉红色电气石晶体矿石
Pink Tourmaline Ore

• 电气石
Tourmaline

辰砂

辰砂又称"丹砂""朱砂",是一种棕红色半透明的矿物晶体,主要成分是硫化汞。辰砂晶体表面具有金属光泽,晶形为板状或者柱状,硬度较小,密度较大,只出产在低温热液的矿床中,常与辉锑矿共生。辰砂是提炼汞的主要的矿物原料。

中国是世界上出产辰砂最多的国家之一。主要产地是贵州东部和湖南西部。古代产出的辰砂多数运到湖南西部的辰溪,再转销全国各地,辰溪成为当时辰砂主要的集散地,"辰砂"一名也由此而来。在古代,辰砂晶体还曾是文人的案头

● 辰砂和白云石共生矿 (图片提供:FOTOE)
Intergrowth of Cinnabar and Dolomite

is its color. Red is the best, followed by blue, green, yellow green and light blue, black tourmaline is relatively poor.

Cinnabars

Cinnabar, also known as "Red Cinnabar", is a brown-red and semi-transparent mineral crystalloid, whose main chemical ingredient is mercuric sulfide. Cinnabar crystalloid has a metal-luster surface, panel or column shape and low hardness grade but high density grade. Cinnabar can only be generated in low-temperature hydrotherm deposits, and often co-exists with stibnite ores. Cinnabar is the primary mineral raw material for extracting mercury.

China is one of the countries producing the most cinnabars in the world. In China, cinnabars are mainly produced in eastern Guizhou and western Hunan. In ancient times, most cinnabars were first shipped to Chenxi City in western Hunan Province, from where it was re-distributed to other regions of China. Chenxi City became the major cinnabar market. The term "Chensha" was named after this city. Cinnabars were popular among scholars and literati in ancient China. In his book *On Exotic Stones*, Zhang Yingwen, a scholar of

文玩。明代文人张应文曾在《论异石》中记载："近时砚山（笔架）书镇（镇纸）有以大块辰砂、石青为者，雅甚。"

the Ming Dynasty, recorded that "It has become an elegant vogue in recent years that scholars like to use big cinnabars to make penholders and paper-weights."

作为颜料和药物的辰砂

中国古人从很久以前就开始用辰砂作为颜料，称为"朱砂"，其粉末呈鲜红色，且不易褪色。"涂朱甲骨"指的就是把辰砂磨成红色粉末涂嵌在甲骨文的刻痕中，醒目且保留时间长。"朱笔御批"指的是封建社会的历代帝王利用辰砂的红色粉末书写的批文，目的是看着醒目和长期保存。印泥是中国人钤印图章用的红色油质颜料，一般也由朱砂制成。作为药物的辰砂一般称为"丹砂"。在传统中医中药理论中，辰砂可入药。

- 朱砂印泥
  Cinnabar Inkpad

Cinnabar Used as Pigments and Medicine

Long ago, the Chinese people started to use cinnabars as pigments, which were then called Zhusha. Powdered zhusha is bright red and does not fade easily. The idiom "Smearing zhusha on oracle bones" means to grind cinnabars into powder and smear it in the snicks of oracle bones, so that the words are eye-catching and lasting. The term "Approval by the Emperor with a Red Brush" means that the approval of the documents by emperors in feudal dynasties in China was written with red cinnabar powder ink, so that the documents were both eye-catching and enduring. Inkpad is an oil-based red pigment that the Chinese people use to seal documents. Inkpad is always made from cinnabar powders. In traditional Chinese medicine theory, cinnabar can be used as medicine and is usually called *Dansha*.

辉锑矿

辉锑矿属于锑的硫化物矿物，颜色呈铅灰色，表面具有金属光泽。单晶呈长柱状或针状，柱面有明显的纵纹；晶体集合体常呈放射

Stibnites

Stibnite belongs to the antimonic sulfide family. It is usually in lead-gray color with a metallic luster. A single crystalloid takes the shape of a cylinder or a needle and there are obvious vertical veins on the sides. The aggregations are in radial or clustered shape. The monocrystal's size varies from several centimeters to several decimeters.

Stibnite is widely distributed in China. The tin mine in Hunan Province is the largest and most famous stibnite production site in the world. Many stibnite deposits have also been found in Guizhou, Guangxi, Shaanxi and other provinces and regions. Stibnite has a strong metallic luster, especially

• 辉锑矿晶簇
Stibnite Druses

状或束状，单晶大小从几厘米到几十厘米不等。

辉锑矿的分布很广，湖南省的锡矿山是世界上最大最著名的辉锑矿产地，贵州、广西、陕西等省区也有不少辉锑矿床产出。辉锑矿具有耀眼的强金属光泽，特别是在解理面或晶面上，可呈现非常灿烂的光泽。明显的晶面纵纹及千姿百态的造型，特别是形如宝剑晶头，如削的长柱形单晶，亮光闪闪，十分惹人喜爱，具有很高的观赏和收藏价值。

雌黄

雌黄属单斜晶系的硫化物矿物。晶体为柱状，常呈叶片状、粒状、放射状、葡萄状、肾状及球状

- **辉锑矿**
  Stibnite

- **雌黄晶簇**
  Orpiment Druses

on the cleavage and crystal sides. Stibnite also has clear vertical veins and diverse shapes. The column-shaped monocrystals, with their shining sword-like heads, are indeed very eye-catching. No wonder they enjoy high values for appreciation and collection.

Orpiment

Orpiment belongs to the monoclinic sulfide family. While orpiment crystals are in a cylinder shape, orpiment minerals often appear in shapes such as leaves, particles, radial, grapes, kidneys, grain and lumps. It is semi-transparent with yellow and light brown colors. Gains on orpiment are of the same color shades,

块体，半透明；颜色呈黄色，微带浅褐色；其条痕与矿物本色相同，但色彩更为鲜明，具有金属光泽。

雌黄主要为低温热液矿床中的典型矿物，雌黄经常与雄黄共生，还见于热泉沉积物和火山凝华物中。中国的湖南石门与慈利交界之处为著名的雌黄产地。

only more vivid with a metallic luster.

As a typical mineral generated in low-temperature hydrotherm deposits, orpiment often co-exists with realgar, and also appears in thermal spring sediments and volcanic lava. The border area between Shimen City and Cili City in Hunan Province is a famous production site for orpiment.

信口雌黄

"信口雌黄"这句成语出自《晋阳秋》，说的是晋朝人王衍喜欢清谈，经常约人闲聊。他手里执拂尘，侃侃而谈，所讲内容经常前后矛盾，漏洞百出。有人质疑时，他就任意更改，随心所欲。人们说他是"信口雌黄"，用来比喻不顾事实，随口乱说。但是，这与"雌黄"这样一个矿物名字又有什么关系呢？原来在中国古代，人们写字时用的是黄纸，如果字写错了，用雌黄的粉末涂一涂，就可以重写了。所以，"雌黄"就用来指代涂改、改写，进而引申为罔顾事实，任意乱讲。

A Mouthful of Orpiment (Making Irresponsible Remarks)

The Chinese idiom "A mouthful of orpiment" was quoted from an ancient book *Stories of Jinyang City*. The story was about a Jin person named Wang Yan, who was very passionate about idle talk. He often invited people to discuss these issues with him. Waving a horsetail whisk, he would articulate his views on Daoism in eloquence. However, his views were often self-contradicting with many loopholes. When someone challenged his views, he would change his opinion randomly and freely. So, local people regarded him as a person with "A mouthful of orpiment" implying his irresponsible remarks. Then, what is the correlation between this story and the mineral "orpiment"? The answer is that in ancient times, people used to write on yellow paper. If there were mistakes, they would smear some yellow orpiment powder on the paper to cover up the mistake, and rewrote on it. In time "orpiment" became the synonym for erasing and rewriting. In time, it was further extended to imply irresponsible remarks based on ignorance of the facts.

雄黄

雄黄晶体属单斜晶系的硫化物矿物,又名"鸡冠石"。单晶体呈不规则块状,大小不一,短柱状,一般以粒状或块状集合体产出。雌黄具有极其艳丽的颜色和光泽,粗大晶体十分稀少;常与透明的方解石、橙黄色的雌黄共生,更显得绚丽多姿。雄黄常呈橘红色,条痕呈淡橘红色,与辰砂相似。

中国湖南慈利和石门交界的

Realgar

Realgar, also known as "Cockscomb Stone", is a member of the monoclinic sulfide mineral family. Its monocrystals are irregular lumps, cylinders or particles. Realgar crystals have flamboyant colors and lusters and are of varying sizes. They often co-exist with transparent calcite and orange colored orpiment. Similar to cinnabar, realgar often shows orange-red color with light orange-red color veins.

Jiepaiyu, located at the border area between Cili City and Shimen City in Hunan Province, is the largest production area of realgar in the world. The largest piece of realgar crystalloid was found in Shimen City, Hunan Province. This 8 centimeters long, 5.4 centimeters wide,

● 雄黄晶簇
Realgar Druse

● 雄黄矿石
Realgar Ore

界牌峪为世界雄黄产地之最。我国发现的最大雄黄晶体，产于湖南石门，长8厘米，宽5.4厘米，高3.5厘米，重255.3克，为世界罕见，现收藏于北京大学地质陈列馆。

3.5 centimeters high realgar crystalloid piece weighs 255.3 grams, very rarely seen in the world. It is now kept in the Geology Pavilion of Peking University in China.

端午节与雄黄

端午节是中国具有几千年历史的传统节日，在每年农历的五月初五。端午时节正值春夏之交，也意味着蚊蝇等害虫进入滋生高峰期。因此在这一天，人们除了吃粽子、赛龙舟以外，还有一个重要的活动就是利用雄黄来防疫消灾，以祛除蚊蝇虫蛇。民间有些地方还将雄黄酒喷洒在房屋壁角阴暗处，或者以雄黄为主加入白芷、薰衣草等香料做成香袋，佩带在腰间。

Dragon-boat Festival and Realgar

In China, the Dragon Boat Festival has a history of several thousand years. Falling on the fifth day of the fifth month according to the lunar calendar, the festival marks the turning point from spring to summer, when all kinds of harmful insects such as mosquitoes and flies are entering into reproductive booms. So, in addition to having glutinous dumpling food and engaging in dragon-boat competitions on this day, people also engage in an important activity on that day, which is to use realgar for epidemic prevention and driving away snakes, mosquitoes, flies and pests. Some people also spray the realgar wine in their house corners, or mix realgar with angelica and lavender to make scent bags and gird them on waists.

绿柱石

绿柱石又称为"绿宝石",是一种环状结构的硅酸盐矿物,属六方晶系。晶体常呈六方柱状,柱面上有纵纹;无色透明的绿柱石较为少见,一般多呈各种色调的浅绿色,具有玻璃光泽。透明、无杂质、无裂纹的绿柱石是宝石级矿物,而一般不透明的绿柱石柱体可作观赏石,直径最粗可达20厘米。

成分中富含铯的绿柱石呈粉红

Beryls

Beryl, also known as "Emerald", is a ring-structured silicate mineral, belonging to the hexagonal crystal family. Its crystals are hexagonal cylinders with vertical veins on the surfaces. As achromatic and transparent beryl stones are very rare, most of them show various shades of green with a glass luster. Transparent and pure beryl stones with no crannies are of precious gem quality and the more commonly seen opaque beryl stones can be used for appreciation purposes. Some of the beryl cylinders are very thick with diameters up to 20 centimeters.

Beryls containing cesium ingredients are called "rose beryl", showing pink color. Beryls containing chromium ingredients are called "Emerald",

• 绿柱石矿
Beryl Ore

• 祖母绿宝石
Emerald

色，称为"玫瑰绿柱石"；含铬的绿柱石呈鲜艳的翠绿色，称为"祖母绿"；含二价铁（Fe^{2+}）时，呈淡蓝色，称为"海蓝宝石"；含少量三价铁（Fe^{3+}）时呈黄色，称为"黄绿宝石"。色泽美丽的绿柱石则是宝石原料，其中尤以祖母绿及海蓝宝石最珍贵。

中国新疆阿尔泰地区是绿柱石的主要产地，甘肃、云南等省也有产出。

赤铁矿

赤铁矿，俗称"红铁矿"，是一种铁的氧化物。赤铁矿有完美的金属闪光菱面体晶体，而在更多

showing bright green color. Beryls containing Fe^{2+} ingredients are called "Aquamarine", showing light blue color. Beryls containing a small quantity of Fe^{3+} ingredients are called "Yellow Green Gem", showing yellow color. Beryls with beautiful luster are raw materials for gems, among them emerald and aquamarine are most precious.

Beryl is mainly produced in A'ertai area, Xinjiang Autonomous Region, China. Some are also produced in Gansu and Yunnan provinces.

Hematites

Hematite, commonly known as "Red Iron Ore", is a member of the ferric oxide family. Hematite has perfect rhombus

- 镜铁矿与水晶共生矿
Intergrowth of Spiegeleisen and Crystal

- 镜铁矿
Spiegeleisen

的情况下，晶体常常是扁平的，更有的形成薄板状，有的样品板状成簇组成玫瑰花状，叫"铁玫瑰"；有时呈鳞片状集合体，称为"镜铁矿"。所有这些结晶很好的赤铁矿变种都是黑色的，但条痕即矿物粉末的颜色都是红色的，所谓肾状铁矿就是这种红色。肾状铁矿是一些放射状的集合体，有肾状的表面。

黄铁矿

黄铁矿的化学成分是二硫化铁，因其浅黄铜的颜色和明亮的金

• 长石、文石、黄铁矿晶钵
Feldspar, Aragonite and Pyrite Crystalloid

crystals with a metallic luster. More often than that, the crystals are in flat or a thin sheet shape. Some hematites with rose-shaped druses are called "Iron Rose". Others with squama-shaped aggregations are called "Spiegeleisen Mineral". All these hematite variations with good crystals are black, but their veins or mineral powders are red. The so-called "kidney shaped ore" appoints to this red color. Kidney-shaped ore is the radioactive aggregation with a kidney-like surface.

Pyrites

The chemical ingredient of pyrite is ferrous disulfide. Because of its light-copper color and bright metallic luster, pyrite is often erroneously identified as gold. So it is also called the "Fool's Gold". As the most widely distributed sulfide mineral, pyrite can be found in many kinds of rocks and stones. Pyrites with beautiful crystalloid shapes have values for appreciation and are deeply loved by the Chinese people. China is one of the major pyrite stores in the world. Famous production areas include Yingde and Yunfu Cities in Guangdong Province, Ma'anshan City in Anhui Province and Baiyin City in Gansu Province.

属光泽，常被误认为是黄金，因此又称为"愚人金"。黄铁矿是分布最广泛的硫化物矿物，在各类岩石中都可出现。晶体形态较好的黄铁矿颇有观赏价值，深受人们喜爱。中国黄铁矿的储量居世界前列，著名产地有广东英德和云浮、安徽马鞍山、甘肃白银等。

石膏

石膏的化学成分是含水的硫酸钙，有次生脉状和湖相沉积两种矿床类型，大都为呈白色的半透明

● **方解石、黄铁矿晶体**
Calcite and Pyrite Crystalloid

● **透石膏晶体**
完全透明的石膏晶体称为"透石膏"，表面具有玻璃光泽，解理面有珍珠光泽，具有较高的观赏和收藏价值。
Transparent Gypsum Crystalloid
The fully transparent gypsum crystals are called "Transparent Gypsums", with glass-like luster on the surface and pearl-like luster on cleavages. Transparent gypsum has high values for appreciation and collection.

Gypsum

The chemical ingredient of gypsum is hydrous calcium sulphate. Generated in two types of mineral deposits of secondary veins and lacustrine sediments, most gypsum appears in the form of white semi-transparent crystal, which can be impressed with fingernails. Hence it is called the "soft gypsum". When soft gypsum loses the crystal water, it becomes "hard gypsum". Gypsum belongs to the monoclinic family. Its

晶体，用手指甲能刻动，所以又称"软石膏"，失去结晶水的称"硬石膏"。石膏属单斜晶系，晶体为板状，呈平行双面、斜方柱等单形。晶面常见纵纹，双晶普遍，常见燕尾双晶，恰如燕子尾部的形态特点。

作为观赏石的石膏有晶体石和造型石两类。晶体石膏在湖北、湖南、贵州、四川、新疆、内蒙古、山东等地均有产出。造型石膏石，产于现代盐湖中。内蒙古戈壁地区、锡林郭勒盟的现代盐湖中有石花状、石柱状和形态各异的造型石膏石产出。

• 石膏晶体
Gypsum Crystalloid

plate-like crystals are in parallel or rhombic cylinder shapes. Gypsum has vertical veins and dual crystals. The commonly seen ones are swallowtail-shaped dual crystals.

Appreciation gypsum entails two types of stones: crystals and modeling stones. Gypsum crystals are produced in Hubei, Hunan, Guizhou, Sichuan, Xinjiang, Inner Mongolia, Shandong and other provinces and regions, and modeling gypsum stones are produced in modern salt lakes. Flower-shaped, cylinder-shaped and diverse-shaped gypsum stones have been produced from modern salt lakes in the Gebi desert area, Xilinguole Prefecture, Inner Mongolia Autonomous Region.

Native Gold

Native gold is a natural mineral with gold as the key ingredient. Its crystals are in the dendrite, particle or squama shapes. Some irregular lumps weigh several tens of kilograms. Native gold possesses a metallic luster, with bright golden color and veins. In China gold mines are mainly distributed in the northeast and northwest regions, as well as Shandong and Hunan provinces.

自然金

自然金是以金为主的自然元素矿物，其结晶一般呈树枝状、粒状或鳞片状，偶见不规则大块体，个别重达数十千克。具金属光泽，颜色和条痕均为光亮的金黄色。中国金矿主要分布在东北、西北、山东、湖南等地。

自然银

自然银，是自然界天然形成的银，常含有一定的杂质。自然银完整的单晶体极为少见，一般晶体往往向一个方向延伸，并发生扭转或挠曲。集合体呈树枝状、不规则薄片状、粒状或块状；新鲜断口呈银白色，但表面往往呈现灰黑色；条痕银白色；具有金属光泽。

Native Silver

Native silver is naturally generated silver containing some impurities. Complete native silver monocrystals are rarely seen. Most of the crystals grow in one direction with twists and turns. Crystalloid aggregations are in dendrite, irregular sheet, particle or lump shapes. Fresh fractures show silver white color with dark graysurface. All the veins show silver white color and metallic luster.

- 与铅锌矿共生的自然银 (图片提供: FOTOE)
Intergrowth of Native Silver and Lead Zinc Ore

> 古生物化石

化石，泛指各种古生物遗体和它们的生活遗迹经过自然界作用保存在地层中变成的石头。这些生物多数为茎、叶、贝壳、骨骼的坚硬部分，经过矿物填充和交替作用，形成仅保留原状、结构及印模的钙化、碳化、硅化的生物遗体、遗物和印痕，具有很高的科研价值、观赏价值和经济价值。

古生物化石可分为实体化石和遗迹化石。前者是指由古生物遗体本身形成的化石，如无脊椎动物的外壳、脊椎动物的骨骼、植物的木质纤维等；后者是指古生物生活、活动时留在沉积地层表面或内部的痕迹或遗物，如动物足迹、卵生动物的蛋化石、古人类所使用的石器等。

> Paleontological Fossils

In the broad sense, fossils refer to stones transformed from paleontological remains stored in the earth strata through the effects of nature. Most of these organisms were stems, leaves, shells and hard parts of the bones. Through the filling by minerals and alternating functions, these organisms became biological remains and relics, retaining only the original form, structure and calcified, carbonized, silicified prints. They enjoy high scientific, ornamental and economic values.

Paleontological fossils can be divided into real fossils and trace fossils. While the former refers to the remains of the organism itself, such as shell invertebrates, vertebrate bones, plants, wood fibers, etc., the latter refers to traces or relics such as animal footprints,

珊瑚石

珊瑚是腔肠动物门、珊瑚纲的无脊椎动物。珊瑚的外胚层细胞分泌石灰质物质，形成珊瑚虫身体的组成部分——外骨骼。人们所见到的珊瑚就是珊瑚虫死后留下的骨骼。而珊瑚化石是古代珊瑚虫的石灰质骨骼经过石化作用保存下来的化石，比珊瑚坚硬，常保存于石灰岩及泥灰岩中。珊瑚化石分布在寒武纪第四纪的地层内，种类多，分

fossilized eggs of oviparous animals, stone tools used by ancient humans, left in the surface or inside the sedimentary stratum of the earth during their activities.

Coral Fossils

Corals are invertebrates under the coral class, coelenterate phylum. The ectoderm cells of coral secrete calcareous material form an integral part of the coral body — the exoskeleton. What people normally see are skeletons of corals after their death. Coral fossils of ancient calcareous

- 圆圆（白珊瑚石）
  The Round (White Coral Fossil)

布广，生存时限短，是划分地质历史的重要参照物。

珊瑚化石种类丰富，可分为四射珊瑚、六射珊瑚、床板珊瑚和日射珊瑚四类。这些化石千姿百态，有盘状、树枝状、拖鞋状、锥状、多角状、圆筒状等，颇得奇石收藏爱好者的青睐。

在中国的古籍中，有许多关于珊瑚石的记载。明代文人林有麟的《素园石谱》中就曾提到珊瑚石。清代文人谷应泰在《博物要览》中记载："珊瑚生海底，作枝柯状，明润

coral skeletons preserved through the petrifaction process become more solid and are often preserved in limestones and marls. Coral fossils are often found in strata of the Quaternary of the Cambrian Period. As corals have many sub-classes and are widely distributed, plus their short existence period, they are often used as a major reference for defining geological history.

Coral fossils can be broadly divided into four categories: four radiant coral, six radiant coral, bedboard coral and sunshine coral. These fossils are deeply adored by collectors of all age groups

● 生命之源（红珊瑚石）
Origins of Life (Red Coral Fossil)

如红玉，中多有空……以高而鲜红者值钱。"可见珊瑚石作为观赏奇石，在当时的市场上已不鲜见。

because of their diverse shapes and postures such as coiled, dendrite, sandal-shaped, conical, multiple angular, cylindrical, etc.

In ancient Chinese books, there are many records of corals. In his book *Scholars' Rocks in Ancient China*, Ming Dynasty literati Lin Youlin mentioned corals. Gu Yingtai, a Qing Dynasty literati wrote in his book *Museum Highlights* the following lines: "Corals grow on sea beds, with branches and a smooth luster like that of the red jade, with many hollows... tall and bright red color corals are more valuable." It can be extrapolated that coral was already a well-known appreciation stone in the then market.

- 月宫仙子（珊瑚石）

中国古代神话传说中，月亮是个宫殿，是嫦娥的住处，月宫中还有一只玉兔常年在月桂树下捣药。这件珊瑚石形状颇似白兔，故而取名"月宫仙子"。

Fairy in the Moon Palace (Coral Fossil)

According to an ancient Chinese legend, there is a palace on the moon where the fairy lady Chang'e lives in. There is also a rabbit in the moon palace who pounds herbal medicine under a laurel tree all year long. Because of the shape of this coral mimicking very much the rabbit, so the stone was named the "moon fairy".

石燕

　　石燕是生活在3.3亿年前的一种海底无脊椎动物，属于腕足类。因其外形类似展翅而飞的燕子，所以被古人称作"石燕"。中国的石燕的产地很多，大多集中在南方。其中，记载最早也最有名的是今天湖南省的祁阳县、零陵县，此处所出石燕曾作为贡品进奉朝廷。

　　石燕个体大小和形状与中小号的菱角相似。由背、腹两壳组成，复壳呈鸟嘴状尖喙，中央有凹下的腹中槽；全壳都布满自喙散出的壳线，两壳相连的绞合线直而长。石燕常单个风化散落山野，也有的多个化石簇生一团，颇具观赏价值。

Spirifer Sinensis (Stone Swallow)

As a member of the brachiopods family, Spirifer Sinensis (Stone Swallows) are invertebrates that lived in deep sea some 330 million years ago. Because its shape is like a flying swallow with spreading wings, it was metaphorically named by the Chinese people in ancient times as "Stone Swallow". Spirifer Sinensis is produced in many parts of China, mostly in the south. Among the production areas, Qiyang and Lingling counties, Hunan Province, produce the best Spirifer Sinensis and have the earliest record of it. They used Spirifer Sinensis as the tribute to the imperial administration.

　　The size and shape of Spirifer Sinensis is similar to that of a small to

• 石燕
Spirifer Sinensis (Stone Swallow)

鸮头贝化石

鸮头贝化石是腕足类化石中个体最大的，其中肥厚型的品种体径常超过10厘米，因形体很像鹰头而得名。其背壳、腹壳双凸，呈心形至球形；腹壳壳喙高突、尖锐，向内弯曲呈钩状；当外壳新鲜未风化时，可见细的同心纹，通常壳面光滑，见不到纹饰。鸮头贝化石主要产于云南东部及广西中部的泥盆纪泥灰岩中。

● 鸮头贝化石 （图片提供：FOTOE）
Owl Head Shell Fossils

medium size water chestnut. Composed of a dorsal and a ventral shell on the back and the belly, Spirifer Sinensis has a beak-shaped mouth and a sunken groove at the center of the belly; the shells are covered with lines radiating from the beak; strands between the two shells are straight and long. Weathered Spirifer Sinensis often scatter in mountains and fields individually. Some fossils are formed into clusters, which make the fossils all the more valuable for appreciation.

Owl Head Shell Fossils

Owl head shell fossils are the largest brachiopod shell fossils, with the size of hypertrophic varieties exceeding 10 centimeters in diameter. It got its name because its shape looks like an owl's head. Both its back and belly shells are bulging and in a heart shape; the beak on the ventral shell is sharp and curved forming an inward hook; before the shell was weathered, fine concentric lines are visible. Usually, the shell surface is smooth with no visible patterns. Owl head shell fossils are found mainly in Devonian marl in eastern Yunnan and central Guangxi.

震旦角石

震旦角石是一种海生无脊椎软体动物化石，是4.4亿年前生活在海洋中的食肉动物。由于沧海桑田，大自然地质历程变迁，今埋藏于坚硬的石灰岩中，故取之不易，得之更难，从而使其更显珍贵。震旦角石形似宝塔、竹笋，故民间俗称"宝塔石""竹笋石""发财石"。常作为贵重礼品馈赠亲朋好友。同时，它还具有很好的学术价值，备受学术界的青睐。

Aurora Cornerstones

Aurora Cornerstone is the fossil of marine invertebrate mollusks which lived some 440 million years ago as an ocean predator. Because of the natural geological processes and changes in the environment, Aurora Cornerstones are now buried in hard limestone. They are difficult to excavate, even more difficult to obtain, making them all the more precious and valuable. Aurora Cornerstones are also called "Pagoda Stones", "Bamboo Shoot Stones", and "Fortune Stones", because of their shapes which look like pagodas and bamboo shoots. Aurora Cornerstones are often used as precious gifts to friends and family Aurora Cornerstones also have good academic values, and are highly admired by the academic community.

● 震旦角石（图片提供：FOTOE）
Aurora Cornerstone

菊石

菊石是已灭绝的海生无脊椎动物的化石，属于软体动物门头足纲，最早出现在古生代泥盆纪初期，繁盛于中生代，广泛分布于世界各地的三叠纪海洋中，白垩纪末期绝迹。现代海洋中的鹦鹉螺是它的近亲。菊石化石的纵剖面呈美丽的螺旋形，棕黄色半透明，色如琥

Ammonites

Ammonites are fossils of extinguished marine invertebrates, under the cephalopods division of the mollusca category. Ammonites first appeared during the early Devonian Period of the Paleozoic Era, and flourished during the Mesozoic Era when they were widely distributed in the Triassic Period seas around the world. They became extinct at

• 菊石
Ammonites

• 菊石

菊石的壳体是一个以碳酸钙为主要成分的锥形管。壳管的始端细小，通常呈球形或桶形，称为胎壳。绝大多数菊石的壳体以胎壳为中心在一个平面内旋卷，少数壳体呈直壳、螺卷或其他不规则形状。

Ammonites

Ammonite shell is a conical tube composed of mainly calcium carbonate. The starting end of the conical tube is narrow, usually in a ball or cylinder shape, known as the casing shell. The majority of ammonite shells scroll up on a flat plane, a few are of straight, spiral or other irregular shapes.

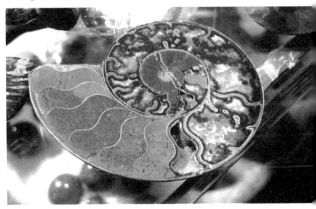

珀，人们喜欢收藏摆放在家中。

海百合化石

　　海百合是一种始见于石炭纪的棘皮动物，生活于海里，具多条腕足，身体呈花状，表面有石灰质的壳。海百合的身体有一个像植物茎一样的柄，柄上端羽状的东西是它们的触手，也叫腕。海百合在死亡以后，这些钙质茎、萼很容易保存下来成为化石，不仅为地质历史时

the end of the Cretaceous Period. Nautilus found in modern oceans is the close relative of Ammonites. The longitudinal section of Ammonites shows a beautiful spiral shape, in a translucent brown color like that of the amber. People collect and display Ammonites in their house.

Crinoid Fossils

Crinoids are a member of the echinoderms family first seen during the Carboniferous Period. They lived in the sea, and had

● 海百合化石 （图片提供：FOTOE）
Crinoid Fossils

期的古环境研究提供重要的证据，也逐渐成为化石收藏家的珍品。

中国的海百合化石多产于贵州省北部的一小块地区，是距今2.3亿年前的中生代三叠纪的遗迹。那时候的贵州地区还是一片深海，后来随着地壳的运动，海底抬升，海百合保存在灰岩中成为化石。灰岩多呈铁灰色或土黄色，而海百合化石为灰白色，二者反差很大，因而海百合图案显得十分清晰而逼真，具有很强的立体感。风格各异的海百合化石成为受人欢迎的观赏石，给人以神秘的美感。

三叶虫化石

三叶虫是一种海生的远古动物，距今5.6亿年前的寒武纪就出现，5亿至4.3亿年前发展到高峰，至2.4亿年前的二叠纪完全灭绝，前后在地球上生存了3.2亿多年。在漫长的时间长河中，它们演化出繁多的种类，有的长达70厘米，有的只有2毫米。其虫体的外壳纵分为一个中轴和两个侧叶，故名"三叶虫"。

中国人发现和收藏三叶虫化石的历史悠久。宋、明、清时代即有

multiple brachiopods. Its body is in the shape of a flower, with calcareous shell on the surface. Part of the crinoid's body is a handle-like stem, on top of which feathery tentacles grow. These tentacles are also called wrists. When crinoids die, the calcareous stem and calyx are easily preserved and transformed into fossils. These fossils serve as important evidence for environmental studies on geological history and they also became fossil collectors' much pursued treasure.

Most of China's crinoid fossils are found in the northern part of Guizhou Province, a small area that hosts the remains of the Mesozoic Age Triassic Period some 230 million years ago. At that time, the Guizhou area was under deep seawater. Later with the movement of the earth crust, the sea bed rose. Crinoids preserved in limestones became fossils. Lime-Stones are mostly iron-gray or khaki in color while crinoid fossils are bright gray. The big contrast in colors makes the crinoid patterns very clear and lively, achieving a strong three-dimensional effect. These diversified crinoid fossils have become popular ornamental stones with a mysterious beauty.

人把玩三叶虫化石，称之为"多福石"，这在宋人笔记和明代曹昭的《格古要论》中都有详细的记述。因三叶虫体形近似飞燕，古人又称之为"燕子石"。三叶虫化石色泽古雅，质地温润，纹彩特异，富有天然之趣，而且它记录了大自然的沧桑变迁，耐人寻味。燕子石还常被制作成砚台、镇纸、笔架、印泥盒等文房用品，以及屏风、花瓶、扇面等装饰品，造型古朴，格调高雅，形艺结合，具有独特风韵。

Trilobite Fossils

Trilobites are ancient marine animals, that emerged about 560 million years ago during the Cambrian Period, developed to its peak between 500 million - 430 million years ago, and became extinct 240 million years ago during the Permian Period. Trilobites existed on earth for a total of 320 million years. In the long course of evolution, they developed into diverse forms. Some were as long as 70 centimeters, others were only 2 millimeters. The outer shell had a central axis and two petals on each side, hence it got the name "Trilobite."

Collecting trilobite fossils has a long history in China. During the Song, Ming and Qing dynasties, many people developed the hobby of fondling trilobite fossils, which was then named the "Happiness Stone". This was noted in the journal of a Song scholar and Cao Zhao of the Ming Dynasty. In this three-volume monograph *Appraisal of Cultural Relics*, Cao Zhao presented a detailed account of the trilobite fossil. Because of the similarity of its shape with that of a swallow, trilobite fossils were also called the "Swallow Stone" in ancient times. Trilobite fossils have elegant colors,

• 三叶虫化石
Trilobite Fossils

鱼化石

鱼类的化石是较为常见的动物化石,主要保存在砂岩、粉砂岩中。早期的原始鱼类以及晚古生代和三叠纪的低等鱼类的化石相对较少,到了侏罗纪,真骨鱼类即高等硬骨鱼类开始出现并迅速繁盛起来,所以硬骨鱼化石在中国许多地区都有出产。

关于鱼化石的形成,宋代杜绾在其所著的《云林石谱》中做了科学的解释:"古之陂泽,鱼生其中,因山颓塞,岁久,土凝为石所致。""因山颓塞"指的是火山爆发,"土"是火山灰。火山爆发,鱼体被火山灰压入湖底,窒息而死,但是由于火山灰的覆盖,鱼体与空气隔绝,不会

gentle textures and unusual patterns. They are full of natural interests. In addition, the fossils record the vicissitudes of nature and induce imagination. Swallow stones have been made into ink stones, paperweights, penholders, inkpad holders and other stationeries, as well as screens, vases, fans and other ornamental items. With their simple shape, elegant style, excellent artisanship, these stones are impregnated with a unique charm.

Fish Fossils

Preserved mainly in sandstones and siltstones, fish fossils are more commonly seen in fauna fossils. Fish fossils of early primitive fish and low-level fish during the Late Paleozoic Era and Triassic Period are relatively small in quantity. Entering the Jurassic Period, teleost fish, advanced bone fish emerged and quickly prospered. This explains why teleost fish

- 辽西鲟鱼化石
  Sturgen Fossils from Western Liaoning Province

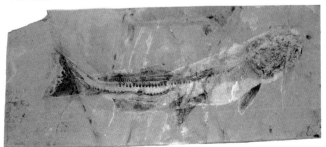

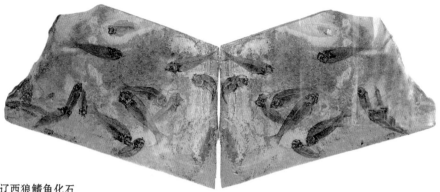

- **辽西狼鳍鱼化石**

  狼鳍鱼是东亚地区有特色的一种远古鱼类，体型较小，头部高度与身体高度相近，大眼睛，背鳍靠后，常常单个或成群地保存在岩层中的同一石板上，神态自然，栩栩如生。在中国，尤其在辽宁以及河北等地相当集中。在侏罗纪晚期，辽宁西部地区曾是一片海洋和湖泊，生活着很多鱼类、两栖类及水生爬行动物。后来由于突如其来的地质变化和火山喷发，大量狼鳍鱼被火山灰埋葬，逐渐形成了化石。

  Lycoptera Fossils from Western Liaoning Province

  Lycoptera is an ancient fish of special characteristics in East Asia. The fish is small with similar size head and body, big eyes and dorsal fins on the back. A single or hoards of lycoptera are often found in the same stone of the rock strata, with natural postures, vivid and lively. In China, especially in Liaoning and Hebei provinces, there are quite some concentrations of lycoptera fossils. In the Late Jurassic Period when western Liaoning was under deep ocean and lake waters, lots of fish, amphibians and aquatic reptiles lived in this area. Later, sudden geological changes and volcanic eruptions buried a large number of lycoptera in the volcanic ash. They gradually transformed into fossils.

腐烂，久而久之，就印刻在火山灰上了。古人早就对鱼化石情有独钟，清代文人沈心在《怪石录》中介绍，在山东莱阳一带，有人把鱼化石制成石屏风出售。

贵州龙化石

　　贵州龙主要生活于三叠纪中期的滨海环境，是地球上最原始的爬行动物，属海生爬行动物的蜥

fossils can be found in many regions in China.

　　Du Wan of the Song Dynasty provided a scientific explanation on the formation of fish fossils in his book *Yunlin Stone Collection* stating: In ancient times, fish used to live in deep lakes. When the volcano erupted, fish were buried under volcanic ash and suffocated to death. Because they were isolated from air, the dead fish body didn't decay. Over time, the shape of the

鳍类。因它的化石在中国贵州省发现，所以被命名为"贵州龙"。贵州龙头近长三角形，眼眶大而圆，四肢细长，脚短且没有变成鳍；其体长最长的有30多厘米，最短的有10厘米左右。贵州龙化石骨骼为纯黑色，骨质致密，呈立体的骨条。

贵州龙化石大多保存较为完整，形象生动，尤其是四只爪子往往显得十分犀利，实属史前爬行动物化石的珍品。在一块块贵州龙化石上，栩栩如生地再现了地质事件突发时的情景，在濒死的一刻，有的贵州龙缩成一团，有的张牙舞

fish got engraved on volcanic ash". The Chinese people have long been fond of ancient fish fossils. In his book *Catalogue of Bizarre Stones* Qing Dynasty literati Shen Xin introduced that in the Laiyang area of Shandong Province, people make screens with fish fossils for sale.

Guizhou Dragon Fossils

Guizhou dragon lived mainly during the mid-Triassic Period when the coastal environment prevailed. As the most primitive reptile on Earth, Guizhou dragon belongs to the lizard finfish branch of the marine reptile family. It was named as Guizhou dragon because

● 贵州龙化石
Guizhou Dragon Fossils

爪，有的垂死挣扎，也有的茫然不知……种种画面引起人们的遐思。

恐龙蛋化石

恐龙是生活在距今约2.3亿年至6700万年前的陆栖爬行动物，灭绝于中生代白垩纪。恐龙是卵生动物，恐龙的卵经过长期地壳演变形成的恐龙蛋化石是一种十分珍贵的古蛋化石，兼具科研、收藏和鉴赏价值。恐龙蛋化石的大小悬殊，小的与鸭蛋差不多，直径不足10厘

● 网格蛋化石
这是一窝5枚较为完整的恐龙蛋化石，产自河南省淅川县，年代为距今约9000万年的晚白垩纪。
Grid Egg Fossils
This is a host of five relatively complete dinosaur egg fossils, found in Xichuan County, Henan Province, formed during the Cretaceous Period about 90 million years ago.

the fossil was found there. Guizhou dragon's head is a long triangle, with large round eye sockets, and four slender limbs. Its feet are short and have not yet evolved into fins; the dragon's body varies between 30 centimeters as the longest and 10 centimeters as the shortest. Petrified Guizhou dragon bones are pure black, with high density, shown as three-dimensional bone stripes.

Most of the Guizhou dragon fossils are well-preserved, and have vivid images, especially the four claws that are often very sharp. Indeed they are treasures of prehistoric reptile fossils. Each fossil records a scene when a sudden deadly geological event took place. Just before the dying moment, some of the dragons curled up, others rattled and ran amuck, some struggled for life, still others got totally lost … all the scenes induce endless imagination.

Dinosaur Egg Fossils

Dinosaurs are terrestrial reptiles that lived between 67 million-230 million years ago, and became extinct during the Mesozoic Cretaceous Age. Being an oviparous animal, dinosaur eggs were transformed into fossils over a

米；大者直径超过50厘米。蛋壳的外表面光滑或有点状和线状的纹路。目前全世界已发现恐龙蛋化石的地点达百余处，而中国是发现恐龙蛋化石最多的国家之一，分布在河南、广东、湖北、山东、江苏、内蒙古、江西等地。其中，河南省的西峡县更是由于恐龙蛋化石的发现而声名远播。当地发现的恐龙蛋化石不仅原始状态保存较好，而且数量多、类型全，其形态包括圆形、扁圆形、长形、橄榄形等。

long period of crustal evolution and became very precious ancient egg fossils, which bear significant values both for scientific research and for collection and appreciation. The size of dinosaur egg fossils varies tremendously. While the small ones are only the size of a duck egg measuring less than 10 centimeters in diameter, the large ones measure more than 50 centimeters in diameter. The surface of the eggshells is smooth or with dots and lines. Currently, dinosaur egg fossils have been found in more than a hundred sites in the world, and China is one of the countries hosting the most sites, including Henan, Guangdong, Hubei, Shandong, Jiangsu, Inner Mongolia, Jiangxi and other provinces and regions. Among them, Xixia County in Henan Province became well-known for dinosaur egg fossils discovered locally. These roundish, oval, oblong and olive-shaped egg fossils were not only well preserved, but also in large quantities with many varieties.

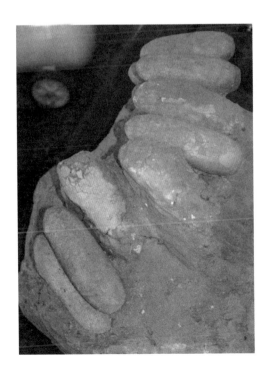

- 恐龙蛋化石
  Dinosaur Egg Fossils

硅化木

　　硅化木又称"树化石",产生于1亿多年前的白垩纪时期,是一种因受到硅化作用而形成的树木化石。这些化石清晰地保存了树木原有的纤维、年轮、节疤、树瘿等。

　　硅化木的化学成分是二氧化硅,含有少量有机质及铁、钙、磷等,常见颜色有浅黄色、黄褐色、红色、棕色、黑色、灰白色等;抛光面具玻璃光泽,半透明至微透

Silicified Wood

Silicified wood, also known as "tree fossils ", was formed during the Cretaceous Age more than one hundred million years ago. It is the fossil of trees formed through the petrification process of trees. These fossils clearly preserved the original tree fiber, rings, knots and galls, etc.

　　The chemical composition of silicified wood is silicon dioxide, containing a small amount of organic matter and iron, calcium, phosphorus, etc. The common colors of silicified wood include light yellow, brown, red, brown, black and gray; polished surfaces of silicified wood that are translucent or slightly transparent bear the luster of glass. In general, silicified wood can be divided into three categories: original mountain stone, wind-eroded stone and water-washed stone.

　　Original mountain stone: Original mountain stones are those silicified trees remaining at the original site and embedded in rocks or buried in the sand. Such stones maintain the original shape and texture of the trees, with clear rings and grains. They look more like tree stubs.

● 横竖大千(树化石)
The World (Tree fossil)

明。通常硅化木可分为三大类：山原石、风砺石和水冲石。

山原石，即树木硅化后还在原生地，嵌在岩层中，或埋在沙土里，保存了原始的形状和质地，年

Original mountain stones of silicified wood were deeply respected and admired by literati during the Tang and Song dynasties. They became one of the traditional ornamental and appreciation stones. Water-washed stones and wind-eroded stones have been recognized by the stone community in recent years, and are gradually becoming new favors in the family of ornamental stones.

• 吉祥（树化玉）

树化玉是一种特殊的硅化木。在漫长的地质过程中，在温压的不断变化中，硅化木继续发生变质作用，重新结晶，主要成分转换为蛋白石玉髓，就形成了树化玉。树化玉的形成需要极其苛刻的地质环境条件。

Auspicious (Jade Tree)

Jade tree is a special member of the silicified wood family. In the long geological process, the constantly changing temperatures and pressures enabled silicified wood to continue with its metamorphism and re-crystallization processes, with the main component converting into opal chalcedony, hence forming a jade tree. The formation of a Jade tree requires extremely strict conditions of the geological environment.

• 树化石
Tree Fossil

轮、脉络清晰，犹如大树残桩。

硅化木山原石在唐宋时期就受到文人的推崇，成为著名的传统观赏石之一，而水冲石、风砺石则在当代才被石界发现和认识，并逐渐成为观赏石家族中的新宠。

风砺石，一般是在水冲石的基础上，经地壳抬升到高原荒漠上，经风沙磨砺而成的硅化木。

水冲石，即树木硅化后经过地壳运动，被洪水、冰川等搬迁至低洼处，再经激流沙石冲撞磨砺，变成质坚、形美、色艳、纹细、皮润、光亮的硅化木。

琥珀

琥珀又称"树脂化石"，是半透明非晶体质的有机宝石，是地质历史上的树木分泌物——树脂经硬化作用形成的。树脂在地壳构造运动中埋入地下后经过千万年的地质变化、挥发、散失、固化而成，树脂中如果包裹有昆虫，则形成昆虫琥珀。当含有琥珀的煤层遭受风化破碎，并经水流搬运、沉积，琥珀可富集成砂矿，例如滨海砂矿。同时，琥珀也产于黏土层和沙砾盐中。

Wind-eroded stone: In general, wind-eroded stones are water-washed stones uplifted to deserts and plateaus through crust movements, and tempered by wind and sand and became silicified wood.

Water-washed stone: Water-washed stones are silicified trees, first moved by flood and icebergs to lower-lying areas during the tectonic movement of the earth crust, and then tempered by currents and debris, finally became silicified wood with hard texture, beautiful shape, bright color, fine grains and smooth surface.

Amber

Amber, also known as "Resin Fossils", is a semi-transparent amorphous organic gem. It was formed through the hardening effect of tree secretions in geological history. Resins were buried underground during the tectonic movement of the earth crust and after millions of years of geological changes, evaporation and losses, resins solidified. If insects are trapped in the resin, insect amber is formed. When the coal bed layer with amber deposits is broken as a result of weathering and is transported by water and sedimentation, amber can be integrated into placers, such as those coastal placers. Amber can also be found in clays and gravel salt.

- 琥珀
  Amber

琥珀依色泽与透明度及纯净度不同，可分为明珀、蜡珀、花珀、水珀等。据古生物地史和煤田地质资料，琥珀形成于石炭纪至第三纪，距今约3.5亿年至几千万年前。琥珀产于沉积地层中、煤系地层中及海滨沙矿中。我国古代把琥珀称为"兽魄""遗玉"等，早在新石器时代的遗址中就出土了琥珀雕刻的装饰物，历经商周秦汉，与古代玉器的发展形影相随。由于琥珀的来源稀少，故非常珍贵。

Amber is divided into several broad categories including Clear Amber, Wax Amber, Flower Amber and Water Amber, based on their color, transparency and purity. According to paleontological history and geological data of coal beds, amber was formed between the Carboniferous and the Tertiary Age, about 350 million years to tens of millions of years ago. Amber can be found in the sedimentary strata, coal-bearing strata and coastal placers. In ancient times, amber was also called "Beast Soul" and "Lost Jade" in China. Amber ornaments were found in the ruins of early Neolithic times. Amber developed at a par with that of jade during the ensuing Zhou, Qin and Han dynasties. Because of its scarcity, amber is very precious.

• 琥珀
Amber